Life Table Techniques
and Their Applications

This is a volume in

STUDIES IN POPULATION

Under the Editorship of: H. H. WINSBOROUGH

Department of Sociology
University of Wisconsin
Madison, Wisconsin

A complete list of titles in this series is available from the publisher on request

LIFE TABLE TECHNIQUES
AND THEIR APPLICATIONS

Narayanan   *Krishnan Namboodiri*
DEPARTMENT OF SOCIOLOGY
THE OHIO STATE UNIVERSITY
COLUMBUS, OHIO

C. M. Suchindran
DEPARTMENT OF BIOSTATISTICS
SCHOOL OF PUBLIC HEALTH
UNIVERSITY OF NORTH CAROLINA
CHAPEL HILL, NORTH CAROLINA

 1987

ACADEMIC PRESS, INC.
Harcourt Brace Jovanovich, Publishers
Orlando   San Diego   New York   Austin
Boston   London   Sydney   Tokyo   Toronto

ACADEMIC PRESS, INC.
Orlando, Florida 32887

United Kingdom Edition published by
ACADEMIC PRESS INC. (LONDON) LTD.
24–28 Oval Road, London NW1 7DX

Library of Congress Cataloging in Publication Data

Namboodiri, N. Krishnan (Narayanan Krishnan), Date
    Life table techniques and their applications.

    (Studies in population)
    Bibliography: p.
    Includes index.
    1. Mortality—Tables. I. Suchindran, C. M.
II. Title. III. Series.
HB1322.N36 1986   304.6'4'0212   86-17213
ISBN 0—12—513930—6 (alk. paper)

PRINTED IN THE UNITED STATES OF AMERICA

87 88 89 90   9 8 7 6 5 4 3 2 1

A000004759805

To the Loving Memory of Unni

Contents

vii

Chapter 5   Statistical Comparison of Life Tables

Chapter 6   Multiple-Decrement Life Tables I

Chapter 7   Multiple-Decrement Life Tables II: Analysis of Follow-up Data

Chapter 8   Multiple-Decrement Life Tables III: General Theory

Chapter 9   Multistate Life Tables

Preface

The development of the subject matter covered in this book has a long history, beginning with the publication in 1666 of John Graunt's "Natural and Political Observations. . . upon the Bills of Mortality." Recently there has been a surge of interest in the topic, particularly in the statistical analysis of factors affecting lifetime, typically focusing on a nonnegative, right-censored response variable exhibiting appreciable variation. We regard the present work as an introduction to this vast and still growing literature. More specifically, in writing this book we have kept in mind demographers, epidemiologists, and others who are interested in descriptive or explanatory analyses of duration of occupancy of a state, e.g., being in school. In describing the time pattern of exit from a state, one gives attention to different ways of exit, e.g., a person enrolled in school may die, drop out, or graduate. The tendency to reenter a state, if permissible, is also of concern in descriptive analyses. In explanatory analyses, one may be interested in identifying factors affecting the length of occupancy of a state (the waiting time to move to another state), e.g., the waiting time to conceive after discontinuation of contraception.

Chapters 1, 2, and 3 are concerned with the traditional life table focusing on death (exit from life). Departing from the usual treatment, attention is given in Chapter 3 to viewing the life table as a Markov process. This departure has been prompted by a desire to prepare the reader for Chapter 9, in which multistate life tables are discussed from that perspective. Chapter 4 is concerned with life tables based on survey and observational data, while Chapter 5 is devoted to comparisons of life tables. These two topics seldom receive attention in textbooks on demographic techniques. They are, however, given a prominent place in treatises on survival analysis. Chapters 6, 7, and 8 are devoted to multiple-decrement life tables, dealing as they do with two or more ways of exit from a given state. Chapter 9, as mentioned, is concerned with multistate life tables, traditionally known by the term increment–decrement life tables. Chapters 10, 11, 12, and 13 deal with explanatory analysis of lifetimes. In Chapter 10, lifetime is viewed as a random variable, and Chapters 11, 12, and 13 cover various approaches to examining factors affecting the "length of life."

In a few chapters, e.g., 3, 10, and 12, it is assumed that the reader is familiar with elementary calculus. Exposure to the basic ideas of matrix algebra is required to follow portions of Chapters 3 and 9. The background necessary to follow the material of Chapters 11 and 12 includes introductory notions regarding regression analysis. In portions of Chapter 11 techniques of categorical data (contingency table) analysis are used. In general, mathematical treatments are kept at an informal level throughout. No attempt has been made, for example, to rigorously prove any assertions. Those who wish to gain deeper insight into the technical details are urged to consult the references cited. The bibliographic notes appended to the various chapters may be of particular help in this connection. Chapter-end "Problems and Complements" often cover matters not addressed in the text. Many exercises are of the paper–pencil variety, but some call for analysis of data using available computer programs.

Both of us were first introduced to life table techniques, in a graduate program in statistics, from the viewpoint of actuarial approaches to the study of mortality. Random variation was seldom mentioned in that connection. Over time, both of us, along with the rest of the profession of demographers, became increasingly accustomed to viewing the life table from the stochastic perspective. Unfortunately, this development has not yet found its way into textbooks on demographic techniques. We hope that the present work partially fills that lacuna.

Portions of the book have been used in graduate courses in demographic techniques and seminars on life table methodology at the University of North Carolina at Chapel Hill and at the Ohio State University.

Chapters 1–5 and 10–13 were first drafted by Namboodiri and 6–9 by Suchindran. Some chapters went through several rounds of revision. It cannot be said that the final product bears much resemblance to the initial version.

Work on this book was started when Namboodiri was on the faculty of the Department of Sociology at the Chapel Hill Campus of the University of North Carolina. Although no specific support was received from any source by either of us for writing this book, we acknowledge with thanks general support received by Suchindran from the Carolina Population Center and the Department of Biostatistics of the University of North Carolina at Chapel Hill, and by Namboodiri from the Department of Sociology of the Ohio State University, and the Department of Sociology and the Carolina Population Center of the University of North Carolina at Chapel Hill.

Chapter 1 | Introduction

The life table (also referred to as the mortality table) is a statistical device used by actuaries, demographers, public health workers, and many others to present the mortality experience of a population aggregate in a form that permits answering questions such as the following: What is the probability that a man aged 30 years will survive to his retirement age, say 70, or that his wife aged 28 now will outlive him? What is the average number of years of life remaining for women who have just reached their 25th birthday? What fraction of babies born in the year 1980 will still be alive in the year 2000? How many of them will survive till their 70th birthday? How many will die before reaching their first birthday?

The life table method is applicable to the analysis of not only mortality but of many measurable processes involving attrition or accession to aggregate size. Thus, the accession to the rank of the married through marriage and remarriage and the attrition to their rank through death, widowhood, or divorce can be depicted in terms of a life table. Similarly, the entrance into the labor force and exit therefrom because of death, temporary withdrawal, or retirement; the personnel turnover in organizations; the loss of production time to workers in a factory because of illness or disability; the age pattern of completing formal education; and many other phenomena can be studied using the life table method. The applicability of the method is by no means confined to human populations. One can apply it to bees and wasps, to mice and dogs, or to any other form of life. The method can be applied to nonliving things, too, for example, to describe the life and death history of automobiles or refrigerators manufactured in a given year by a given company or to study the length of life of light bulbs, washing machines, airplanes, and of many others.

A very wide area of application of the life table method is the analysis of survival data derived from laboratory studies of animals or from clinical studies of humans who have acute diseases or are fitted with artificial devices (e.g., IUD, pacemaker). Survival data may include survival time, response to a given treatment, or patient characteristics related to response and survival. The end point need not be death; it can be a first response, relapse, or development of some prespecified trait. One of the special features of survival data is that usually such data are censored. Different kinds of censoring are possible. To illustrate, suppose ten rats have been exposed to carcinogens by injecting tumor cells into their footpads. The time taken for a tumor to develop and reach a given size is observed. But suppose, because of cost constraints, the investigator decides to terminate the experiment at the end of a 30-week period. During this time, suppose only eight rats had developed a tumor of the given size. Then all that the investigator knows about the remaining two rats is that in neither case had the tumor developed or reached the given size within the first 30 weeks. For this reason the observations on these two rats are called censored. In this case since all rats were simultaneously brought under observation, the censoring concerns the right-end of the data. In many clinical studies, however, patients enter the study at different times during a fixed period of study. Suppose for example, the period fixed for a study is from January 1, 1980 to December 31, 1980. Patients may enter the study at any time during the 12-month period involved. Some of them may die before the end of the study; some may be lost to follow-up; and the rest may be alive at the end of the study. For the "lost" patients, survival times are greater than or equal to the interval between the time of their entrance and the time of the last contact. For patients who are still alive at the termination of the study, survival times are at least equal to the interval between their entrance and the end of the study. Both of these last two types of observations are censored, and, since their entry times are not simultaneous, the censoring involved in their cases is somewhat different from that involved in the laboratory study of rats mentioned above.

Data situations of the kind just mentioned as well as various applications of the life table method alluded to earlier will be considered in detail later on in the book. For the moment, however, we turn to an examination of the anatomy of the life table in the simple situation in which there is no censoring and where the only factor to be concerned about is attrition due to a single cause, namely, mortality.

1 The Anatomy of a Mortality Table

Consider a population of hypothetical creatures with a lifespan of 4 years. (Lifespan is defined as the oldest age to which survival is possible.) Suppose 700 infants are born to these creatures in a given year, say year z, and we are

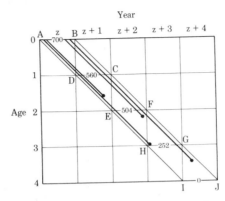

Fig. 1.1. Lexis diagram showing life lines in age–time plane.

able to trace these infants from their birth until the last survivor among them dies. Suppose we find that, of the 700 newborns,

140 die in the first year of life, i.e., before reaching their first birthday,
56 die in the second year of life, i.e., after reaching the first birthday, but before reaching the second,
252 die in the third year, i.e., after completing the first two years of life, but before completing the third, and
252 die in the fourth year, i.e., before reaching the fourth birthday.

It is instructive to diagramatically represent these numbers in what is known as the Lexis diagram (so named after Wilhelm Lexis who introduced it in 1875), shown in Fig. 1.1. Age is laid out on the vertical axis and time on the horizontal, both axes using the same time unit (e.g., one year). Each individual life is represented by a life line inclined at 45 degrees to either axis; the starting point is the time of birth on the horizontal axis, and the end point represents the age and time of death.

The life and death history of the hypothetical body of lives just described, when represented in a Lexis diagram, will have 700 life lines starting from the horizontal axis in the year z. Of these, 140 terminate (die) in the parallelogram ABCD, that is before reaching the horizontal at age 1. Of those lines reaching the horizontal at age 1, 56 terminate (die) in the parallelogram CDEF, that is, between the horizontals at ages 1 and 2. Similarly, 252 life lines terminate in EFGH and the rest in GHIJ. The numbers of life lines that reach (start from) horizontals at successive ages 0, 1, 2, 3, and 4 are respectively 700, 560, 504, 252, and 0.

Now, by relating the number dying between the xth and the $(x + 1)$th birthdays to those who survive to the xth birthday, we get an estimate of the

Table 1.1

Life Table for a Hypothetical Population with a Lifespan of Four Years

Age interval x to $x + n^a$	Probability of dying in the age interval for those alive at the beginning of the interval $_nq_x$	Number alive at the beginning of the interval l_x	Number dying in the interval $_nd_x$	Number of years of life lived in the interval $_nL_x$	Number of years of life remaining T_x	Average number of years of life remaining $\mathring{e}_x$
(1)	(2)	(3)	(4)	(5)	(6)	(7)
0–1	0.2	100,000	20,000	90,000	238,000	2.38
1–2	0.1	80,000	8,000	76,000	148,000	1.85
2–3	0.5	72,000	36,000	54,000	72,000	1.00
3–4	1.0	36,000	36,000	18,000	18,000	0.50

^{a} The age intervals are conventionally "open" at the right. Thus 0 to 1 stands for the interval from birth up to but not including the first birthday. A standard notation for such intervals is $[x, x + 1)$.

probability of dying between the xth and $(x + 1)$th birthdays, given survival till the xth birthday. Thus from the given data, we obtain the following figures:

Age interval x to $x + 1$		Probability of dying in the interval
0	1	$\frac{140}{700} = 0.2$
1	2	$\frac{56}{560} = 0.1$
2	3	$\frac{252}{504} = 0.5$
3	4	$\frac{252}{252} = 1.0$

In general, the probability of dying in the age interval x to $x + n$, denoted by the symbol $_nq_x$ is the basic quantity used in constructing the mortality table (life table) as can be seen from Table 1.1, whose anatomy is the following:

Column 1, x to $x + n$: The period of life between two exact ages, that is, between two birthdays, the xth and the $(x + n)$th. Thus, 1 to 2 in this column means the period between the first and second birthdays.

Column 2, $_nq_x$: The probability that a person who is alive at the beginning of the indicated age interval, i.e., at x, will die before reaching the end $(x + n)$ of the age interval. For example, according to Table 1.1, the probability that a person who reaches his or her second birthday will die before reaching the next birthday is 0.5.

Column 3, l_x: The number alive at (or surviving till) the beginning of the indicated age interval. In practice, when constructing a life table, one usually starts with an arbitrary number of newborns (often a convenient power of 10, such as 1,000, 10,000, or 100,000) and assumes that these newborns will experience attrition due to mortality according to the pattern exhibited by the $_nq_x$ values. This starting number is called the *radix* of the table. In Table 1.1, the radix is 100,000.

Column 4, $_nd_x$: The number of persons in the body of lives in question who die within the indicated age interval x to $x + n$. Thus, according to Table 1.1, 80,000 die in the age interval 1 to 2, that is, that many die after reaching the first birthday but before reaching the second.

Column 5, $_nL_x$: The number of years of life lived by the body of lives in question within the indicated age interval: 90,000 years of life during the first year, 76,000 in the second year, and so on. It is worth noting that in any age interval x to $x + n$, those who survive to the end of the interval will have lived n years each, while each of those who die in the interval will have lived only a fraction of n years. If deaths are uniformly distributed over the interval, one may take each person who died in the interval to have lived on the average $\frac{1}{2}n$ years in the interval. Thus, under this assumption,

$$_nL_x = nl_{x+n} + (n/2)_nd_x$$
$$= nl_{x+n} + (n/2)(l_x - l_{x+n})$$
$$= (n/2)(l_x + l_{x+n})$$

The figures in the $_nL_x$ column in Table 1.1 have been calculated using this formula. Thus, for the first age group, $90,000 = \frac{1}{2}(100,000 + 80,000)$, for the second age interval, $76,000 = \frac{1}{2}(80,000 + 72,000)$, and so on. The assumption that deaths are evenly distributed in a given age interval may not be tenable in some situations for some age intervals. When that is the case, the formula for the calculation of $_nL_x$ becomes somewhat more complicated (see Chapter 2).

Column 6, T_x: The total number of years of life remaining for the body of lives in question after surviving till the beginning of the indicated age interval. Note that this figure can be obtained by adding the $_nL_x$ for the indicated age interval and those for all the subsequent age intervals. Thus in Table 1.1, $T_2 = 72,000 = 54,000 + 18,000$; $T_0 = 238,000 = 90,000 + 76,000 + 54,000 + 18,000$, and so on.

Column 7, $\overset{\circ}{e}_x$: The average number of years of life remaining for the body of lives in question after reaching the beginning of the age interval indicated. Thus according to Table 1.1, those who reach their first birthday have, on the average, 1.85 more years to live; those who survive to their second birthday have on the average 1.00 more year to live, and so on. Note that the figures in this column can be obtained by dividing the figures in the T_x column by the corresponding figures in the l_x column. Thus the first entry is obtained as the ratio $\frac{238,000}{100,000}$, the second as the ratio $\frac{148,000}{80,000}$, and so on.

From the description given above, it should be clear that some of the life table functions can be directly calculated from others. The following formulas summarize the more frequently used interrelations among the life table functions.

$$_nd_x = l_x - l_{x+n}$$

$$_nq_x = {_nd_x}/l_x; \qquad l_{x\,n}q_x = {_nd_x}$$

$$l_{x+n} = l_x(1 - {_nq_x}) = l_x - {_nd_x}$$

$$_nL_x = nl_{x+n} + a{_nd_x}$$

where a is the length of the interval x to $x + n$ lived on the average by those dying in the interval. If $a = n/2$, this last formula becomes

$$_nL_x = nl_{x+n} + (n/2){_nd_x} = nl_{x+n} + (n/2)(l_x - l_{x+n}) = (n/2)(l_x + l_{x+n})$$

The corresponding formula in terms of a is

$$_nL_x = nl_{x+n} + a(l_x - l_{x+n}) = al_x + (n - a)l_{x+n}$$

Once the $_nL_x$ figures have been computed, the T_x figures are obtained by cumulating the $_nL_x$ column from below, and the life expectancies are then obtained by division of T_x by l_x.

A Remark on the Unit of Time

It should be emphasized that the unit of time used in the life table need not always be one year. Months, weeks, days, hours, etc., can serve as appropriate units of time. Furthermore, even when the unit of time used is one year, it is legitimate to consider the number of survivors to exact age $x + t$ or the total or average number of life years remaining for the group of lives under consideration after they have reached exact age $x + t$, where $0 \le t \le 1$.

An Alternative Interpretation of The Life Table Functions

In the description of the life table given above, the frame of reference used was the lifetime mortality experience of a single birth cohort, i.e., of a group of lives born in a given unit of time (year). We shall now give another interpretation to the life table functions. For this purpose imagine the following hypothetical situation. Each year l_0 births occur in a population (where, 10 is the radix of a given life table). Each birth cohort experiences the mortality pattern of the given life table. If this were to continue for a number of years, the population becomes *stationary* in the sense that its size and age composition remain unchanged over time. A census taken in this population at any moment after the stationary character has been attained would count $_nL_x$ persons in the age interval x to $x + n$. We refer to the stationary population as the *life table population* also. According to this framework we have the following interpretations for the life table functions.

l_x The number of persons in a stationary (life table) population who survive till their xth birthday.

$_nd_x$ The number of deaths in a year between ages x and $x + n$ in the stationary (life table) population.

$_nL_x$ The number of persons aged x to $x + n$ in the stationary (life table) population.

T_x The number of persons who are aged x years or more in the stationary (life table) population.

The other functions have the same interpretation as given earlier.

2 Types of Life Tables

Life tables discussed so far may be called *single-decrement* tables, since their construction involves taking into account only one attrition factor (mortality). Contrasted with such tables are *multiple-decrement* tables, in which two or more attrition factors are assumed to operate concurrently. For example, the size of the never-married population dwindles through death or first marriage. Mortality rates may be combined with first-marriage rates to produce a double-decrement first-marriage table.

The q_x values for a specific cause of decrement in a multiple-decrement table may be used in constructing an ordinary single-decrement table to exhibit the age pattern of survival for that cause by itself, that is, when all other causes are eliminated. Thus if we have a double-decrement table for mortality and first marriage, we can obtain one single-decrement table for mortality and one for first marriage.

In multiple-decrement tables as in single-decrement tables there are no increments (new entrants or re-entrants) to the group of lives under study. There are, however, numerous problems in which the group of lives under study is subject to attrition as well as accession (decrements as well as increments). An example occurs in labor-force analysis. The attrition factors in this case include mortality, temporary withdrawal due to disability and similar conditions, and permanent withdrawal due to retirement. New entrants to labor force may join at various ages, and those who withdraw for temporary periods may reenter at the end of the withdrawal periods. Tables that take into account attritions as well as accessions are known as *increment–decrement* tables or *multistate life* tables.

Any of the life tables mentioned so far may be constructed so as to portray the experience of an actual birth cohort or the collection of experiences of various cohorts during a fixed period of time. The life table based on the mortality and survivorship experiences of an actual birth cohort is known as the *generation* or *cohort life* table. To be contrasted with this is the *current life* table, which is based on the mortality rates derived from age-specific death rates observed in a single year or a short period such as three years. The current age-specific death rates are converted into q_x values and are then applied to a hypothetical birth cohort to derive the current life table.

Yet another distinction to note is between *complete life tables* and *abridged life tables*. In a complete life table information is given for every single year of age from birth until the last applicable age. In abridged life tables, information is given only for broader age intervals such as x to $x + 5$ years.

Problems and Complements

1. Show that the reciprocal of the expectation of life at birth is equivalent to the crude death rate of the life table (stationary) population.

2. You are given the values in the life table shown in Table 1.2. Complete the life table.

Table 1.2

Age interval	$_nq_x$	$_nd_x$	$_nL_x$
0–1	0.0569	5,700	95,400
1–2	0.0082		
2–3	0.0063		
3–4	0.0041		
4–5	0.0072		
5–6	0.0155		
6–7	0.1001		
7–8	0.5008		
8–9	0.7093		
9–10	1.0000		

3. Give a general formula to calculate $_5L_x$ and $_5q_x$ for $x = 0, 5, 10, \ldots$, given l_x and $\mathring{e}_x$ values for every fifth age $(0, 5, 10, \ldots)$. Illustrate using the life table constructed in exercise 2.

4. Suppose $[x]$ and $[y]$ are two persons randomly chosen from those who have just reached their xth and yth birthdays, respectively. The mortality pattern prevailing in the population is depicted in a given life table. Use the life table functions to answer the following questions.

a. What is the probability that $[x]$ will still be alive at the $(x + n)$th birthday?

b. What is the probability that both $[x]$ and $[y]$ will still be alive when $[x]$ is exactly $(x + n)$ years of age and $[y]$ exactly $(y + n)$ years of age?

c. Suppose $[x]$ and $[y]$ are married to each other, $[x]$ being the husband and $[y]$ the wife. Suppose further that at the time of their marriage $[x]$ was exactly 25 years of age and $[y]$ was exactly 20 years of age. Give formulas to calculate each of the following:

   i. the probability that both will still be alive 40 years later;

   ii. the probability that the husband will be a widower by that time;

   iii. the probability that the wife will be a widow by that time.

d. Compute the probabilities mentioned in c, Parts i, ii, and iii, using the following l_x values from the Life Table for the United States for the year 1963.

x	White male	White female
20	96,058	97,234
25	95,256	96,928
60	75,342	86,313
65	65,741	80,893

5. A very common demographic application of the life table is in the calculation of the number of persons expected to be alive at future dates given the present population and age-specific mortality rates. A commonly used formula in this connection is

$$K^{(5)}(x + 5, x + 10) = K^{(0)}(x, x + 5)_5 L_{x+5}/_5 L_x$$

where $K^{(t)}(x, y)$ stands for the number of persons in the age interval x to y at time t. The ratio $_n L_{x+t}/_n L_x$ is often called the survival ratio or the survival rate. Express the following in terms of appropriate life table values (functions).

a. The proportion of the population 40 to 44 years of age in year 1980 who will survive 5 years. (Note that age is reckoned in completed years here and throughout this question whenever an interval such as 40 to 44 years is mentioned.)

b. The proportion of the population 75 years and over in 1980 who will live another 10 years.

c. The proportion of the population 53 years of age at the beginning of 1980 who will survive to the end of 1980.

d. The proportion of infants born in the year 1980 who will survive to the end of 1980.

e. The proportion of infants born in 1980 who will survive to their first birthday.

f. The proportion of infants born in 1980 who will survive to the age interval 20 to 24 years.

6. Writing q_x for $_n q_x$ where $n = 1$ and defining $p_x = 1 - q_x$, what does the product $p_0 p_1 \cdots p_{n-1}$ stand for?

7. In a life table the life expectancy at birth and at age 75 are, respectively, 69.89 years and 8.7 years. If $l_0 = 100,000$ and $l_{75} = 48,170$, what proportion of the stationary population will be 75 years or older?

8. In a life table whose radix is 100,000, the life expectancy at birth and at age 75 are 68.23 years and 8.52 years, respectively, and 6 percent of the stationary population is 75 years of age or older. How many in the stationary population survive from birth to exact age 75?

9. Suppose the life lines of each birth cohort in a stationary population are represented in a Lexis diagram. You are also told that births are uniformly distributed over time in the population. Show that under these assumptions, the number of years of life lived by those entering or passing through the square bounded by ages x and $x + 1$ and time points corresponding to the beginning and end dates of any calendar year is L_x (omitting the subscript before L when $n = 1$).

Bibliographic Notes

John Graunt (1666) is regarded as the inventor of the notion of diminution due to death in the size of a group of babies born at a given time point. From general observations he suggested the following pattern: 100 at age 0 diminishing to 64 by age 7, to 25 by age 27, and so on. Graunt's original publication has recently (1964) been reproduced in the *Journal of the Institute of Actuaries* (**90**, pp. 1–61). During the eighteenth century, various attempts were made to construct such age patterns of survival, using numbers of deaths at each age. It was only much later that the convention of using population counts in the estimation of the age pattern of attrition due to death became established. William Farr's (1841) *Fifth Report of the Registrar General* (London) contains a history of early life tables and also a brief account of the construction of one of the first national tables. P. R. Cox's (1975) two-volume monograph *Population Trends* contains an account of the origin, nature, and uses of life tables.

A number of books on survival analysis have appeared recently. Mann *et al.* (1974), Gross and Clark (1975), and Lawless (1982) focus largely on parametric methods of particular distributions. Kalbfleisch and Prentice (1980) deal with the multiplicative hazards model in detail, and Miller (1981) discusses parametric as well as nonparametric methods. Elandt-Johnson and Johnson (1980) cover actuarial and demographic applications, Tuma and Hannan (1984) concentrate on sociological problems, Cox and Oakes (1984) have written for the applied statistician interested in survival analysis, and Lee (1980) has written for biomedical researchers. Mode's (1985) and Manton and Stallard's (1984) monographs are addressed to biomathematicians.

Among recent papers illustrating the use of survival analysis in occupational epidemiology are those of Darby and Reissland (1981) and Breslow *et al.* (1983). For applications in nuptiality analysis, see, e.g., Krishnamoorthy (1979a), Espenshade (1983, 1985), Hofferth (1985a,b), Schoen *et al.* (1985). For application in labor force analysis, see, e.g., Hoem (1977), Schoen and Woodrow (1984), and U.S. Department of Labor, Bureau of Labor Statistics (1982a,b). The papers in the collection edited by Bongaarts *et al.* (1985) illustrate the use of survival analysis in the field of family demography. Manton and Stallard (1984) cover applications in medical demography, and Land and Hough (1985) indicate some applications in the study of voting behavior. Smith (1980) illustrate the application of life table methods to data from World Fertility Survey (WFS) on infant and child mortality, age at marriage, marital dissolution and remarriage, birth and pregnancy intervals, and breastfeeding. Other WFS publications on life table analyses of fertility data include those of Rodriguez and Hobcraft (1980) and Hobcraft and Rodriguez (1980).

Oakes (1981) has provided a review of developments since the publication of Cox's (1972) influential paper on "regression models and life tables." In an expository paper, Peto *et al.* (1977) have described the application of some of the simpler methods in the analysis of clinical trials.

There is an enormous literature pertaining to the area of "life testing" (experiments aimed at collecting, in industrial settings, life-length data so as to estimate such things as "mean time to failure," "failure rate," "survival function," and "reliable life.") Among the well-known works are Proschan and Serfling (1974), Mann *et al.* (1974), Barlow and Singpurwalla (1975), Gross and Clark (1975), Tsokos and Shimi (1977), Bain (1978), and Nelson (1982).

Chapter 2 | Mechanics of Life Table Construction

This chapter is concerned with the mechanics of life table construction. For convenience of exposition we deal with only decrements due to mortality. Other forms of decrements could of course be similarly studied.

In broad terms, the present chapter is organized as follows: First we deal with the complete life table, and then turn to the abridged life table. Under complete life table construction, before discussing how different columns of the life table are to be computed, we discuss various formulas for the computation of the life table mortality rates. In the section on abridged life tables also, a number of methods for calculating the life table mortality rates are discussed.

Throughout this chapter, it is assumed that good quality data on population and deaths are available. Making adjustments for defects in the basic data will not be given any attention. It will be assumed that the basic data are from population censuses and vital statistics registration systems. Data obtained on the basis of sample surveys, clinical observations, and the like will receive attention in a later chapter.

1 The Complete Life Table

As mentioned in Chapter 1, a complete life table gives information for each single-year age interval, i.e., each age interval of unit length (one year) starting from an integer age (e.g., 0, 1, 2, ...). The basic data from which one usually attempts to construct a complete life table are derived from population censuses and vital statistics registration systems, although procedures are available for the construction of complete life tables from census data only or from registration data only, under certain assumptions.

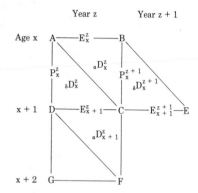

Fig. 2.1. Flow chart to illustrate calculation of exposure to the risk of death.

When census data or vital statistics of an actual population are used in the construction of a life table, it is essential that the reference period (time) as well as the age interval involved be clearly understood. With this in mind we introduce the following notation (see Fig. 2.1).

P_x^z the number of persons who, at the beginning of calendar year z, are between ages x and $x + 1$ (lines crossing AD in Fig. 2.1)

E_x^z the number of persons who, during calendar year z, attain exact age x (lines crossing AB in Fig. 2.1)

D_x^z the number of deaths that occur during calendar year z among persons who have attained age x at last birthday (lines which terminate in ABCD)

$_\alpha D_x^z$ the number of deaths occurring in calendar year z between the xth birthday and the end of the year (lines that terminate in ABC, representing deaths among those who reach their xth birthday during the calendar year z)

$_\delta D_x^z$ the number of deaths occurring in calendar year z between the beginning of year z and the $(x + 1)$th birthday (lines which terminate in ADC, these representing deaths during year z at age x in completed years among those who reached their xth birthday in the year $z - 1$)

It should be clear that $P_x^{z+1} = E_x^z - {}_\alpha D_x^z$, $E_{x+1}^z = P_x^z - {}_\delta D_x^z$, and that $D_x^z = {}_\alpha D_x^z + {}_\delta D_x^z (E_x^z - P_x^{z+1}) + (P_x^z - E_{x+1}^z)$.

Three Formulas for q_x

We now present three different ways in which one may compute q_x from census and vital statistics. (Recall that q_x is the conditional probability of dying in the age interval x to $x + 1$, given survival till age x. Here x is a positive integer.)

One method is to relate deaths in the parallelogram ABEC in Fig. 2.1 to the lines crossing AB:

$$q_x^{z|z+1} = ({}_\alpha D_x^z + {}_\delta D_x^{z+1})/E_x^z \tag{2.1}$$

The disadvantage of this method is that it mixes deaths in two years, z and $z + 1$. (The superscript $z|z + 1$ for q indicates this.)

A second method is to relate deaths in the parallelogram ACFD to the lines crossing AD, giving

$$q_x^z = ({}_\delta D_x^z + {}_\alpha D_{x+1}^z)/P_x^z \tag{2.2}$$

One disadvantage of this method is that it mixes deaths in two age groups, x to $x + 1$ and $x + 1$ to $x + 2$. Another disadvantage is that it is not strictly applicable to the age group 0 to 1. (Why?)

A third method is to focus on deaths in the square ABCD and relate ${}_\alpha D_x^z$ to the lines crossing AB and ${}_\delta D_x^z$ to the lines crossing AD:

$$q_x = 1 - (1 - {}_\alpha D_x^z/E_x^z)(1 - {}_\delta D_x^z/P_x^z) \tag{2.3}$$

If we are willing to make the assumption that the probability that a person who enters a calendar year after the xth birthday but before the $(x + 1)$th will die before the end of the year remains constant from one year to the next, at least in the short run, we can derive Eq. (2.3) from Eq. (2.1). For from (2.1)

$$q_x^z = (E_x^z - E_{x+1}^{z+1})/E_x^z = 1 - E_{x+1}^{z+1}/E_x^z$$
$$= 1 - (E_{x+1}^{z+1}/P_x^{z+1})(P_x^{z+1}/E_x^z)$$
$$= 1 - (1 - {}_\delta D_x^{z+1}/P_x^{z+1})(1 - {}_\alpha D_x^z/E_x^z) \tag{2.4}$$

Now, if the probability that a person who enters a calendar year after the xth birthday and before the $(x + 1)$th will die before the end of the year is constant from one year to the next, then

$${}_\delta D_x^{z+1}/P_x^{z+1} = {}_\delta D_x^z/P_x^z \tag{2.5}$$

Substitution in (2.4) from (2.5) gives (2.3).

Separation Factors

In order to apply any of the formulas given above, one needs to have D_x^z separated into ${}_\alpha D_x^z$ and ${}_\delta D_x^z$. Often published data fail to provide this information. Hence the question arises as to how to make this separation under such circumstances. Suppose we assume that, for each x, the fraction $f_x^z = {}_\delta D_x^z / D_x^z$ remains invariant in the short run and over populations with more or less the same overall mortality level, so that once they are calculated for different x's based on detailed tabulation of deaths for one population at one time period, they can be used for other populations or for the same population at other times. Then given D_x^z, we estimate ${}_\delta D_x^z$ by taking the product of f_x^z with D_x^z. The term *separation factors* is applied to the fractions f_x^z.

The tabulation ideal for direct computation of the separation factors obviously is the one that separates D_x^z into ${}_\alpha D_x^z$ and ${}_\delta D_x^z$. Tabulations in which the same unit of time (e.g., year, month, day) is consistently used in indicating the time period of birth and of the death of the deceased, as well as to measure the age completed by the deceased at the time of death, and which classify deaths occurring in each time period by age at death and the time period of birth of the deceased provide the information needed to calculate the separation factors. The following tabulation of deaths in calendar year z illustrate such a compilation:

Age in completed years	Year of birth	
	$z - \text{age}$	$z - \text{age} - 1$
0	${}_\alpha D_0^z$	${}_\delta D_0^z$
1	${}_\alpha D_1^z$	${}_\delta D_1^z$
2	${}_\alpha D_2^z$	${}_\delta D_2^z$

Some countries such as Germany provide this type of tabulation, but many countries do not. When tabulations of deaths according to calendar year of birth are not available, the best one can do is to obtain approximate values for the separation factors, the degree of approximation attainable being a function of the nature of the data available. To illustrate, suppose infant deaths occurring in each quarter of year z are classified by age measured in terms of quarter years. The available data may be represented diagrammatically as shown in Fig. 2.2.

We know the number of deaths in each square. Clearly, deaths falling above the dotted squares relate to births of year z, and those falling below the dotted squares relate to births of year $(z - 1)$. Those falling in the dotted squares are not easily assigned. If we are willing to assume, however, that these deaths are equally likely to belong to the two birth cohorts z and $z - 1$, then we have a rule needed to separate D_0^z into ${}_\alpha D_0^z$ and ${}_\delta D_0^z$. The rule is: assign all deaths in the full squares below the diagonal in Fig. 2.2 plus one-half of the deaths in the diagonal (dotted) squares to ${}_\delta D_0^z$, and the rest to ${}_\alpha D_0^z$. The procedure

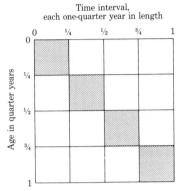

Time interval,
each one-quarter year in length

Fig. 2.2. Flow chart to illustrate calculation of separation factors when data on time of birth and age at death are available in quarter years.

illustrated above could be extended to tabulations involving time-periods such as months, weeks, days, etc., with no difficulty.

It may happen that infant deaths are available for each quarter (or for each month) of year z but not classified by age (in quarter years or months). Or infant deaths may be classified by age, but not by the time period (quarter or month of the year) of occurrence. In such cases, the deaths occurring in each time period or age will have to be distributed over age intervals or time intervals according to some rule, such as even distribution. The procedure illustrated above may then be applied to obtain the separation factors.

Values of separation factors calculated by these methods have shown that there is a strong positive relationship between the separation factor for the age interval 0 to 1 and the overall mortality level in the population, especially the overall infant mortality level. Shryock and Siegel (1973, p. 414) suggest the following pattern for the separation factor for under one year of age:

Infant mortality rate ($= 1000q_0$)	200	150	100	50	25	15
Separation factor for age 0–1	0.40	0.30	0.25	0.20	0.15	0.05

It may be noted that small inaccuracies in the separation factors do not materially affect the estimate of q_x. In most modern populations, it would be an acceptable practice to use 0.5 as the separation factor for ages above one year, especially for ages above 5 years. [Consistent with the assumption regarding the stability of separation factors, we shall use the notation f_x instead of f_x^z for separation factors from now on.]

Estimation of q_x for $x < 5$ years

The method one uses for computing q_x for the very young ages is dictated by the type and quality of data available. Because underenumeration of young children is a common feature of censuses, the usual tendency is to rely on birth and death registration data, when they are available and are of good quality, for the calculation of q_x values for the young ages. Even when census data are of acceptable quality, the tendency is to rely on registration data for the calculation of q_x for very young ages because more information is usually available in registration data than in the census, making it possible to pay special attention to the distinct peculiarities of mortality during these ages.

Suppose that by using registration data or otherwise one is able to trace birth cohorts and represent their death and survival experiences in a Lexis diagram (see, e.g., Fig. 2.1). Then one may use formula (2.4) to compute q_x for ages under 5 years. These formulas are applicable even when there is some degree of underregistration of vital events, provided that the degree of underregistration is the same for births as for deaths.

If sufficient details are available in vital statistics tabulations, it is possible to use quarter-years, months, or some other time period as units in the construction of data sets for the computation of q_x for young ages. England and Wales, for example, uses quarter-years for this purpose. To see what is involved in using such time units, let $^tE_0^z$ denote the number of births in the tth quarter of calendar year z (see Fig. 2.3). Suppose we want to calculate the probability that an infant born in calendar year z will die before completing the first quarter-year of life. Based on the information available in Fig. 2.3, we could estimate this probability in any of the following four ways (using separation factor equal to 0.5):

$$^1q_0^z = \frac{\text{deaths in square B}}{(^4E_0^{z-1} + {}^1E_0^z)/2} \tag{2.6a}$$

$$^1q_0^z = \frac{\text{deaths in square C}}{(^1E_0^z + {}^2E_0^z)/2} \tag{2.6b}$$

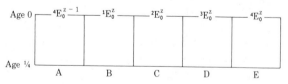

Fig. 2.3. Flow chart to illustrate the use of data in quarter years in the estimation of probability of dying at young ages.

$$^1q_0^z = \frac{\text{deaths in square D}}{(^2E_0^z + {}^3E_0^z)/2} \tag{2.6c}$$

$$^1q_0^z = \frac{\text{deaths in square E}}{(^3E_0^z + {}^4E_0^z)/2} \tag{2.6d}$$

It is logical to combine the ratios on the right-hand sides of (2.6a)–(2.6d), by weighting each ratio with its own denominator, to obtain a single estimate of the probability of dying in the first quarter-year of life. The estimate thus defined is

$$^1q_0^z = \frac{^1D_0^z}{\frac{1}{2}{}^4E_0^{z-1} + {}^1E_0^z + {}^2E_0^z + {}^3E_0^z + \frac{1}{2}{}^4E_0^z} \tag{2.7}$$

where $^1D_0^z$ stands for the number of deaths in calendar year z among infants in the first quarter year of their life ($^1D_0^z$ = the sum of deaths in squares B, C, D, and E in Fig. 2.3). Similarly we could obtain the following probabilities

$$^2q_0^z = \frac{^2D_0^z}{\frac{1}{2}{}^3E_0^{z-1} + {}^4E_0^{z-1} + {}^1E_0^z + {}^2E_0^z + \frac{1}{2}{}^3E_0^z} \tag{2.8}$$

$$^3q_0^z = \frac{^3D_0^z}{\frac{1}{2}{}^2E_0^{z-1} + {}^3E_0^{z-1} + {}^4E_0^{z-1} + {}^1E_0^z + \frac{1}{2}{}^2E_0^z} \tag{2.9}$$

$$^4q_0^z = \frac{^4D_0^z}{\frac{1}{2}{}^1E_0^{z-1} + {}^2E_0^{z-1} + {}^3E_0^{z-1} + {}^4E_0^{z-1} + \frac{1}{2}{}^1E_0^z} \tag{2.10}$$

where $^iq_0^z$ denotes the probability that an infant survives to the beginning of the ith quarter year of life in the first year and dies before that quarter ends. By adding we have

$$q_0^z = {}^1q_0^z + {}^2q_0^z + {}^3q_0^z + {}^4q_0^z \tag{2.11}$$

as an estimate of the probability of dying in the first year of life.

The procedure just described could obviously be extended to obtain estimates of q_1^z, q_2^z, etc., using death statistics in calendar year z and birth statistics in years, z, $z - 1$, $z - 2$, etc.

Estimation of q_x for the Not Very Young and Not Very Old Ages

The calculation of q_x for single years of age in the main body of the life table, usually from age 5 to age 85, is essentially an exercise in graduation techniques. Several methods are described below. Each of them starts with grouped population and death data, using 5-year age intervals (usually) for grouping.

Let $_5p_x^{z+0.5}$ denote the number of persons who as of the middle of calendar year z have completed their $(x + n)$th birthday ($n = 0, 1, 2, 3, 4$), and let $_5D_x^z$ denote the number of deaths in calendar year z with age at death in completed years being $x, x + 1, x + 2, x + 3$, or $x + 4$.

Method 1. From the population figures in successive age intervals, one obtains the corresponding figures for single-year age intervals by interpolation. Similarly, from the aggregated numbers of deaths, the corresponding numbers of deaths for single-year age intervals are obtained. From these quantities q_x^z is estimated using formula (2.12).

$$q_x^z = \frac{D_x^z}{\frac{1}{2}(P_x^{z+1} + P_x^z) + \frac{1}{2}D_x^z} \tag{2.12}$$

The average of the two population figures is usually approximated by the midyear population. This was the technique used in the construction of the United States Life Tables for 1900 and 1910.

Method 2. From the sequence

$$_5P_x^{z+0.5}, \qquad x = 5, 10, \ldots \tag{2.13}$$

one obtains the corresponding figures for pivotal ages 7, 12, ..., using King's formulas. The values thus obtained are called pivotal values. King's formulas for obtaining pivotal values are

$$Y_7 = 0.192 \, _5Y_5 + 0.016 \, _5Y_{10} - 0.008 \, _5Y_{15} \tag{2.14}$$

$$Y_{x+2} = -0.008 \, _5Y_{x-5} + 0.216 \, _5Y_x - 0.008 \, _5Y_{x+5}; \qquad x = 10, 15, \ldots \tag{2.15}$$

from which the required formula for population are obtained by replacing Y by P and those for deaths by replacing Y by D.

From the pivotal values of population and deaths thus computed, one computes q_x^z by applying (2.12), replacing as in Method 1, the average population by the mid-date population. From the q_x values for the pivotal ages, one computes the corresponding values for the intervening ages by the use of the Karup-King third-difference osculatory interpolation formula (see Shryock and Seigel, 1973).

Values of q_x to fill the gap between the highest age in the very young ages for which q_x has been computed from birth and death statistics and the first pivotal age are obtained by interpolation. (One may use for this purpose, graphic interpolation, the Lagrange interpolation formula, or any formula designed to produce smooth junctions.)

Method 3. From the grouped data one obtains the central age-specific death rates:

$$_5M_x^z = {_5D_x^z} / {_5P_x^{z+0.5}} \tag{2.16}$$

From $_5M_x^z$, we calculate $_5q_x^z$ following the procedures for the construction of abridged life tables (see Section 2). From the $_5q_x$ values the corresponding l_x values in the life table are calculated, using successively the relationship

$$l_{x+5} = l_x(1 - {_5q_x})$$

with l_0 arbitrarily chosen. Then the l_x values for the intervening single years of age are obtained by interpolation.

A Brief Comment on Interpolation Formulas

Since the computation of q_x when constructing a complete life table from 5-year age data involves the use of interpolation formulas, it may be worthwhile to make a few remarks about various interpolation formulas commonly used in such exercises. The most commonly used interpolation formulas to compute single-year q_x values are: Karup-King's third-difference osculatory interpolation formula, Shovelton's six-point osculatory formula, Beer's minimized fifth-difference procedure, and Jenkins' modified osculatory fifth-difference formula. The Karup-King formula has the property that it reproduces the pivotal values, thereby insuring conformity with the original data. The other formulas do not have this property as much as the Karup-King formula does. In computing the United States life tables for whites for 1939–1941, the Karup-King formula was used to interpolate for ages 12 to 26 years, where it was considered desirable to retain the irregularities observed in the mortality rates; however, for ages 32 and over q_x at single years were computed using Jenkins' formula on the ground that at these ages the objective of producing a curve smoother than the one provided by the Karup-King formula was to be emphasized. Thus the selection of the interpolation formula to be used in a particular exercise depends mostly on the desired balance between smoothness and closeness of fit to the data. It is impossible to have both smoothness and closeness of fit to the data at the same time; generally, when one of the two improves the other worsens. With the improvement of data quality, analysts have increasingly begun to show a tendency to give priority to closeness of fit to the data.

Estimation of q_x for the Very Old Ages

Population and death data at the very old ages, when they are available, are generally disregarded in computing a life table, mainly because they are considered inaccurate. It has therefore been a common practice to use arbitrary methods for computing q_x at the very old ages (usually 85 and above). For practical purposes, any reasonable method is satisfactory, since the arbitrariness involved in the method has only a small effect on the life table as a whole. The major requirement that is usually kept in mind when choosing a procedure in this connection is that the procedure should produce a smooth junction with the q_x values already computed and a smooth upward progression of q_x with advancing age. Among the various methods employed are the following:

Method 1. Assume that the q_x values increase at the older ages by about the same rate as that found at the upper end (higher ages) of the main body of the life table. Empirically it has been found that a reasonable assumption is that every age at the older ages has a mortality rate 10 percent higher than that at the preceding age.

Method 2. A second method is to fit a cubic curve so as to pass through the last three computed values of q_x and $q_\omega = 1$, where ω is an arbitrarily chosen end of the total lifespan (e.g., $\omega = 110$).

Method 3. A third method is to fit a Gompertz curve ($q_x = Bc^x$) to the end portion of the main body of the life table and use the fitted curve for extrapolation. It has, however, been reported, based on a detailed analysis of Medicare data in the United States, that Gompertz' law does not accurately represent the overall mortality pattern beyond age 90. (See Wilkin, 1981.)

Method 4. Yet another method is to accept the experience of another population for which the data are known to be accurate. Thus, to close the United States 1949–1951 life tables, the mortality experience of Union Civil War veterans observed from mid-1945 to mid-1954 was used. (All but one of the veterans had died by the end of the observation period, and the age data from the group were considered reliable.) See, however, the comment on Method 3 given above. A crucial consideration is whether the cause of death structure is the same in the two populations.

Remark. There is empirical evidence indicating that in industrially advanced populations the rate of increase with age in the force of mortality tapers off around age 90 and that by age 100 it becomes virtually age invariant. With the accumulation of more accurate data on the mortality experiences of the extremely old, increasing attention is being given to this phenomenon.

Derivation of the Remaining Columns of the Life Table

From the interrelationships of the life table functions mentioned in Chapter 1, it should be easy to see that starting with the values of q_x and the arbitrarily chosen l_0 ($= 100,000$, say), the values in the l_x and d_x columns are obtained by repeated application of the relationships: $d_x = l_x q_x$ and $l_{x+1} =$

$l_x - d_x$. To calculate the L_x column, the following formula may be used:

$$L_x = a_x l_x + (1 - a_x) l_{x+1} \tag{2.17}$$

where a_x is the average length of time lived in the interval by those dying in it. Between ages 1 and very old age, a_x is commonly given the value 0.5. In the extreme old age, where mortality rates sharply increase with age, the use of 0.5 for a_x is not theoretically justifiable, but in practice one uses it. Nothing beyond what has already been said in Chapter 1 needs to be said about the calculation of the other columns of the life table.

2 Abridged Life Tables

Abridged life tables, as mentioned in Chapter 1, do not give the life table functions for every single-year age interval. For many practical purposes one may need the life table functions only for every fifth or tenth age or the corresponding age intervals. Moreover, in many instances the time or skill needed to prepare a complete life table may not be available. For these reasons abridged life tables instead of complete life tables are often computed and a number of methods have been developed for their computation. Several of these methods are described below.

A First Approximation

Suppose in a calendar year z, N persons attain their xth birthday. We want to know what proportion of them may be expected to live until their $(x + 5)$th birthday. The information available is the (annualized) central age-specific death rate $_5M_x^z$ and the basic data used in its calculation. From this information, how would one infer the five-year survival probability applicable to the cohort that attained the xth birthday in year z? Without making any further assumptions, it would be impossible to infer the survival probability in question. Different assumptions may lead to different estimates. We start with a simple set of assumptions. Suppose the N persons in question are exposed to a constant rate $(= \mu)$ of decrement due to deaths during the age interval x to $x + 5$. Then the expected number of survivors to their $(x + 5)$th birthday would be $Ne^{-5\mu}$. If we further assume that $_5M_x^z$ is a good estimate of μ, then our estimate of the proportion of the original group who survive to their $(x + 5)$th birthday would be $\exp(-5 {_5}M_x^z)$. Correspondingly, our estimate of the proportion of the original group who die before reaching their $(x + 5)$th birthday would be $1 - \exp(-5 {_5}M_x^z)$, which represents the life table quantity $_5q_x$. This estimate of $_5q_x$ leaves much to be desired, since the assumptions underlying it sound implausible; for example, the constancy in the rate of decrement is unlikely to hold true. Naturally, efforts have been made to improve the estimate.

The Reed–Merrell Method

One of the early attempts to improve the estimate given above for $_5q_x$ has been that of Lowell J. Reed and Margaret Merrell (1939). Reed and Merrell suggested, on empirical grounds, the following improvement:

$$_nq_x = 1 - \exp(-n\,_nM_x - an^3\,_nM_x^2) \qquad (2.18)$$

where a is a constant, n is the length of the age interval, and $_nM_x$ the (annualized) central age-specific death rate for the age interval. Based on a study of the United States life tables computed from the 1910 census data, Reed and Merrell empirically determined that $a = 0.008$ would produce acceptable results. On that basis they prepared several tables from which one could directly read $_nq_x$ corresponding to observed values of $_3M_2$, $_5M_x$, and $_{10}M_x$. Those tables are probably of less use today because of the availability of electronic calculators.

It may be of interest to note that when $n = 5$, with $a = 0.008$, the Reed–Merrell formula assumes the form

$$_5q_x = 1 - \exp[-_5M_x(5 + _5M_x)] \qquad (2.19)$$

which may be simpler for computation on the hand calculator. With $_5M_{70} = 0.05977$, we get 0.26097 as the Reed–Merrell estimate of $_5q_{70}$, as can be easily verified.

Reed and Merrell in their (1939) paper gave special formulas for age intervals 0 to 1, 1 to 2, ..., based on corrections for an assumed degree of underenumeration, an approach that is not general.

As for computing the $_nL_x$ column, the Reed–Merrell method involves using the following formulas:

$$L_0 = 0.276l_0 + 0.724l_1$$
$$L_1 = 0.410l_1 + 0.590l_2$$
$$_4L_1 = 0.034l_0 + 1.184l_1 + 2.782l_4$$
$$_3L_2 = -0.021l_0 + 1.384l_2 + 1.637l_5$$
$$_5L_5 = -0.003l_0 + 2.242l_5 + 2.761l_{10}$$

These formulas were derived by fitting equations of the form

$$_nL_x = Al_0 + Bl_x + Cl_{x+n}$$

where A, B, and C are constants, such that $A + B + C = n$, to the values from 24 United States life tables covering years 1900 to 1930. For the ages beyond 10, the method involves obtaining $_nL_x$ values by differencing the T_x values, which are computed using the following formulas:

$$T_x = -0.20833l_{x-5} + 2.5l_x + 0.20833l_{x+5} + 5\sum l_{x+5a}$$

if the life table uses five-year age intervals, and

$$T_x = 4.166667l_x + 0.833333l_{x+10} + 10\sum l_{x+10a}$$

if the life table uses ten-year age intervals, the summation in both being with respect to a from 1 to the end of the lifespan. These latter formulas were obtained under the following assumption.

The area under the l_x curve between any two ordinates is approximated by the area under a parabola that passes through those two ordinates, and the preceding and succeeding ordinates, if the life table is based on five-year intervals, and the following ordinate only, if the life table uses ten-year intervals.

The Keyfitz–Frauenthal Method

Keyfitz and Frauenthal in a paper published in 1975 suggested the following approximation for $_nq_x$ in terms observed age-specific death rates and population counts:

$$_5q_x = 1 - \exp[-5(_5M_x + C)] \tag{2.20}$$

where

$$C = \frac{(_5P_{x-5} - {}_5P_{x+5})(_5M_{x+5} - {}_5M_{x-5})}{48\,_5P_x}$$

$_nP_x$ being the observed population in the age interval x to $x + n$. The mathematical basis for this approximation is outlined in Chapter 3. (As shown there, the Reed–Merrell formula could be seen as a special case of the Keyfitz–Frauenthal formula.) To illustrate, suppose we are given the following:

x	$_5P_x$	$_5M_x$
60	4,192,000	0.027483
65	3,294,000	0.039958
70	2,330,000	0.059770

To calculate $_5q_{65}$, we first compute

$$C = \frac{(4,192,000 - 2,330,000)(0.059770 - 0.027483)}{48(3,294,000)}$$

$$= 0.0003802$$

This gives

$$_5q_x = 1 - \exp[-5(0.039958 + 0.0003802)] = 0.18265$$

As for the calculation of the $_nL_x$ column, the Keyfitz–Frauenthal method involves the use of the formula

$$_nL_x = \frac{n(l_x - l_{x+n})}{\ln l_x - \ln l_{x+n}}\left[1 + \left(\frac{n}{24}\right)(_nM_{x+n} - _nM_{x-n})\right] \tag{2.21}$$

Greville's Method

Four years after Reed and Merrell published their method for the construction of abridged life tables, Greville (1943) suggested the following formula for the conversion of the observed central age-specific death rates to the corresponding life table mortality rates:

$$_nq_x = \frac{_nM_x}{(1/n) + _nM_x[(1/2) + (n/12)(_nM_x - k)]} \tag{2.22}$$

where k is a constant that may slightly vary from one life table to another but may be taken without incurring any major error to be equal to 0.09. (Greville showed also that his formula given above implies the Reed–Merrell formula, thus supplying a theoretical basis for the Reed–Merrell method.) For $_5M_{65} = 0.039958$, we get by applying (2.22), with $k = 0.09$

$$_5q_{65} = \frac{0.039958}{0.2 + 0.039958[0.5 + (0.416667)(0.039958 - 0.09)]}$$

$$= 0.18234$$

For computing the $_nL_x$ column, Greville suggested that the following method be used. Assume that $_nm_x$, the central age-specific death rate in the life table (stationary) population, is the same as the corresponding observed rate $_nM_x$; that is,

$$_nM_x = _nd_x/_nL_x \tag{2.23}$$

This leads to the formula for estimating $_nL_x$

$$_nL_x = _nd_x/_nM_x \tag{2.24}$$

For the terminal (open) age interval (x, ω), the d_x value is equal to the l_x value, and hence the corresponding L_x value is obtained as the ratio of the l_x to the terminal observed age-specific death rate.

Chiang's Method

Suppose there are N persons in a population who reach their xth birthday in a calendar year z. Let $_nq_x$ be the probability that a person randomly chosen from this group will die before reaching his (her) $(x + n)$th birthday. This means that we would expect $N\,_nq_x$ persons in this group to die within n years after their xth birthday, and the remaining $N(1 - \,_nq_x)$ to survive till their $(x + n)$th birthday. Let us designate by $_na_x$ the fraction of the interval, from the xth to the $(x + n)$th birthday, lived by those dying in the interval, on the average. Then the total number of years of life lived by the group as a whole during the interval between the xth and $(x + n)$th birthdays is the sum of the following two parts:

1. $N(1 - \,_nq_x)n$, which represents the total number of years of life lived by those who survive to the $(x + n)$th birthday, and
2. $N\,_nq_x n\,_na_x$, which represents the total number of years of life lived by those who die in the interval before reaching the $(x + n)$th birthday.

Following the conventional definition of the age-specific death rates we thus have

$$_nm_x = \frac{N\,_nq_x}{N(1 - \,_nq_x)n + N\,_nq_x n\,_na_x} \tag{2.25}$$

which simplifies to

$$_nm_x = \frac{_nq_x}{n - (1 - \,_na_x)n\,_nq_x} \tag{2.26}$$

Solving for $_nq_x$ from (2.26) we obtain

$$_nq_x = \frac{n\,_nm_x}{1 + (1 - \,_na_x)n\,_nm_x} \tag{2.27}$$

Equation (2.27) suggests that $_nq_x$ can be estimated from the observed $_nM_x$ by substituting $_nM_x$ for $_nm_x$ in (2.27), provided, of course, that $_na_x$'s are known. Chiang (1960a) found, on the basis of empirical investigations, that $_na_x$ is more or less invariant with respect to sex, race, cause of death, geographic location, and other demographic variables, and hence concluded that once $_na_x$ is determined for each age group in one population, it could be used for many populations. One set of Chiang's estimates of $_na_x$ is reproduced below:

Age group	$_na_x$	Age group	$_na_x$
0–1	0.09	40–45	0.54
1–5	0.39	45–50	0.54
5–10	0.46	50–55	0.53
10–15	0.54	55–60	0.52
15–20	0.57	60–65	0.52
20–25	0.49	65–70	0.52
25–30	0.50	70–75	0.51
30–35	0.52	75–80	0.51
35–40	0.54	80–85	0.48

Using Chiang's estimate of $_5a_{65} = 0.52$, we obtain $_5q_{65}$ corresponding to $_5M_{65} = 0.039958$ as

$$_5q_{65} = \frac{5(0.039958)}{1 + (1 - 0.52)5(0.039958)} = 0.182307$$

To calculate $_nL_x$, one uses the following formula:

$$_nL_x = nl_{x+n} + n_na_{x\,n}d_x \tag{2.28}$$

the logic of which is simply this: Each person who survives to the $(x + n)$th birthday lives n years during the interval between the xth and the $(x + n)$th birthdays; those who die between the xth and $(x + n)$th birthdays live, on the average, n_na_n years of life between the two birthdays; hence the total number of years of life lived by those reaching the xth birthday between that birthday and the $(x + n)$th is equal to $nl_{x+n} + n_na_{x\,n}d_x$. For the terminal (open) age interval the corresponding quantity is obtained as the ratio of l_x value for the beginning of that interval to the observed age-specific death rate for the interval.

Obviously, the general applicability of this method depends on the generality of the $_na_x$ values given above. If they are not invariant over time and over populations, the method loses much of its attraction. Many more empirical investigations of the kind Chiang refers to in his publications need to be done before one can determine the degree of invariance of the $_na_x$ values.

Method of Reference to a Standard Table

If we write $_ng_x$ in place of $n(1 - {_na_x})$, Eq. (2.27) takes the form

$$_nq_x = n_nm_x/(1 + {_ng_{x\,n}m_x}) \tag{2.29}$$

Solving for $_ng_x$ from (2.29), we obtain

$$_ng_x = n/_nq_x - 1/_nm_x \qquad (2.30)$$

Similarly from (2.28) we obtain

$$_nL_x = nl_{x+n} + (n - _ng_x)_nd_x$$

Now suppose we have access to a life table prepared for a census year and the observed central age-specific death rates upon which the life table is based, and suppose further that we want to construct a life table for a postcensus year, given the following assumptions.

Assumption A. The relation between the observed age-specific death rate in the census year and the corresponding life table mortality rate remains applicable in the postcensus year for the relationship between the observed age-specific death rate in that year and the life table mortality rate to be computed for that year.

Assumption B. So does the relationship between the l_x and the L_x columns of the life tables. (That is, whatever be the relationships between the two columns in the life table for the census year, they hold for the life table for the postcensus year also.)

For convenience of reference, let us refer to the life table of the census year as the standard table. We shall use the conventional symbols for the life table functions of the standard table and the underlying observed age-specific death rates, and shall use an asterisk for the functions to refer to the corresponding quantities in the life table for the postcensus year.

Now we notice that if we insert on the right-hand side of (2.30) $_nq_x$ and $_nm_x$ from the standard life table, we may get one value for $_ng_x$, and if we put instead $_nq_x$ and the observed $_nM_x$ for the census year, we may get another. Let us denote the former by $_ng_x$ and the latter by $_ng'_x$.

In computing the life table for the postcensus year, we may use, by virtue of Assumption A, the following formula:

$$_nq_x^* = n_nM_x^*/(1 + _ng'_{xn}M_x^*) \qquad (2.32)$$

to convert the observed age-specific death rate ($_nM_x^*$) in the postcensus year to the corresponding life table mortality rate ($_nq_x^*$). Also, for the computation of the $_nL_x$ column of the life table, we may use, by virtue of Assumption B, formula

$$_nL_x^* = nl_{x+n}^* + (n - _ng_x)_nd_x^*$$

Abridged life tables for postcensus years in the United States have been prepared by this method.

Problems and Complements

1. Suppose you are able to trace an actual birth cohort in a population and record its death and survival experience. Represent the information in a Lexis diagram and with the help of the diagram express P_x and E_x in terms of E_0, $_\alpha D$'s and $_\delta D$'s.

2. Show that formula (2.3) in the text is equivalent to

$$q_x^z = 1 - (P_x^{z+1}/E_x^z)(E_{x+1}^z/P_x^z)$$

3. Show that an approximation to formula (2.3) is

$$q_x^z \simeq (_\alpha D_x^z/E_x^z)(_\delta D_x^z/P_x^z)$$

4. Show that formula (2.3) is exact even if there is underreporting of births and deaths as long as the underreporting is of the same magnitude in births as in deaths.

5. You are given births registered in 1979, 1980, and 1981. You are also given infant deaths occurring to these birth cohorts, classified by age at death (in completed years). What formulas would you use to compute q_0 applicable to the period 1979–1981?

6. In France as of January 1, 1954, there were 293,238 males aged 50 years old at last birthday. The corresponding figure as of January 1, 1955 was 292,010. Of the 3,081 deaths reported in calendar year 1954 among males aged 50 at last birthday, 1544 occurred to those whose 50th birthday was in 1954 and the rest occurred to those whose 50th birthday was in 1953. Use this information to estimate q_{50} using formula (2.3).

7. In Exercise 6 assume that $_\delta D_{50}^{z+1}/P_{50}^{z+1} = _\delta D_{50}^z/P_{50}^z$ and on that basis estimate q_{50} using formula (2.1).

8. Show that an estimate of $_\delta D_{x+1}^z$ is $(_\delta D_x^z/P_x^z)P_x^{z+1}$ and, hence or otherwise, show that

$$_n q_x = \frac{_\alpha D_x^z}{E_x^z} + \frac{_\delta D_x^z}{P_x^z} - \frac{_\alpha D_x^z}{E_x^z}\frac{_\delta D_x^z}{P_x^z}$$

9. In addition to the assumption made in Exercise 6, assume that

$$P_x^{z+1}/E_x^z = P_x^z/E_x^{z-1}$$

On the basis of these assumptions, derive the following formula:

$$q_x^z = _\alpha D_x^z/E_x^z + _\delta D_x^z/E_x^{z-1}$$

A special case of this formula is

$$q_0^z = _\alpha D_0^z/E_0^z + _\delta D_0^z/E_0^{z-1}$$

or

$$q_0^z = (1 - f)D_0^z/E_0^z + fD_0^z/E_0^{z-1}$$

which relates death among cohort z to that cohort and those among cohort $z - 1$ to that cohort.

10. A complete life table has been defined as the one that gives the number of survivors at successive birthdays and the various life table functions for individual ages or single-year age intervals. It should be noted that the term complete is largely arbitrary; thus, for example, one may construct a life table that gives even more details than the so-called complete life table. Given in the accompanying table are the l_x column of a French life table for the year 1957 for infants under 1 year of age (both sexes combined).

Age interval (days)	l_x	Age interval (days)	l_x
0–30	100,000	180–210	97,613
30–60	98,508	210–240	97,506
60–90	98,249	240–270	97,401
90–120	98,041	270–301	97,311
120–150	97,868	301–332	97,226
150–180	97,729	332–365	97,149

(Source: Pressat (1972), p. 148)

Compute the d_x and the q_x columns.

11. a. Express $_5M_x^z$ as a weighted average of M_{x+t}^z, $t = 0, 1, 2, 3$, and 4.

   b. Express M_x^z as a weighted average of the age-specific death rates for each quarter-year of age in the age interval x to $x + 1$.

   c. Suppose the age interval x to $x + 1$ has been divided into k subintervals each of length $1/h$ of a year: x to $x + h$, $x + h$ to $x + 2h$, and so on. Show that in obvious notation

$$M_x = \frac{\Sigma \,_h P_{x+rh} \,_h M_{x+rh}}{\Sigma \,_h P_{x+rh}}$$

   d. Extend the formula in part c to $_5M_x$.

12. Using the results of Exercise 11 or otherwise discuss the bounds of error in the estimate of the life table mortality rate (Keyfitz, 1977, pp, 43–44).

13. Given a complete life table, describe how you would derive a life table that starts at age 15 rather than at age 0.

14. In constructing life tables for a census year, it is fairly common to compute age-specific death rates using the following formula:

$$\text{age-specific death rate} = \frac{\text{deaths in years } z - 1, z, \text{ and } z + 1}{3 \,(\text{population enumerated in } z)}$$

where z is the census year. What do you think is the rationale for this procedure?

15. In some instances the following formula has been used for the computation of age-specific death rates from death statistics and census data:

$$M_x = \frac{3D_x}{P_{x-1} + P_x + P_{x+1}}$$

What do you think is the rationale for this procedure?

16. Suppose you have calculated q_x for ages 0, 1, 2, 3, and 4, using birth and death registration data. You wish to prepare the main body of the life table in such a way as to reproduce the observed age-specific death rates. Describe an iterative procedure involving the use of $_n g_x$ introduced in the text. [HINT: *Step 1.* Convert the central death rates (for five-year age groups) to life table mortality rates using the formula

$$_5q_x = \frac{5 \,_5M_x}{1 + \,_5g_x \,_5M_x}$$

(initially using 2.5 for $_5g_x$ for all x).

Step 2. Calculate l_x for $x = 5, 10, \ldots$ using the familiar formula $l_x = (1 - {}_5q_x)l_{x-5}$

Step 3. Obtain l_x for intervening ages by interpolation, and compute ${}_nL_x$ using the formula

$$_5L_x = \tfrac{1}{2}l_x + l_{x+1} + l_{x+2} + l_{x+3} + l_{x+4} + \tfrac{1}{2}l_{x+5}$$

Step 4. Compute the 5-year central age-specific death rate for the life table population

$$_5M_x = {}_5d_x/{}_5L_x$$

Step 5. Compute a new ${}_ng_x$ using the formula

$$_5g_x = (5/{}_5q_x) - (1/{}_5m_x)$$

Now go back to Step 1 and repeat the whole process. The iteration stops when two successive iterations produce more or less identical results.]

17. Misstatement of age is common in censuses and death records. For older persons there seems to be a tendency to exaggerate rather than understate age. The net result of this tendency is in general an underestimation of mortality at older ages. If the misstatement of age is of a certain type, methods can be devised to reduce the impact of age incorrectness on mortality estimates. Suppose, for example, there exists an age a such that exaggeration of age is confined to ages a or higher and that the age data below a are not defective. If it is also reasonable to assume that the age pattern of mortality follows the Gompertz law for ages beyond early childhood, then we may fit the Gompertz distribution for the truncated data above childhood but below age a and then extrapolate the curve for estimating mortality beyond age a. Another approach might be to use an appropriate model life table that fits well the mortality pattern till age a to estimate the pattern above age a. A third approach is the following, suggested by Horiuchi and Coale (1982). This method is based on the Gompertzian assumption mentioned above plus the assumption that the age distribution above a is stable. The Horiuchi–Coale formula for estimating life expectancy at age a is

$$[e(a)]^{-1} = M(a^+)\exp\{\beta r(a^+)[M(a^+)]^{-\alpha}\}$$

where $\theta(a)$ is the life expectancy at age a; $M(a^+)$ is the age-specific mortality rate for ages a and above; $r(a^+)$ is the growth rate of the population of age a and above, and α and β are constants whose values, empirically determined, are for given a,

a	50	55	60	65, 75, 85
α	1.0	1.1	1.2	1.4
β	0.2827	0.2068	0.1565	0.0951

[Horiuchi and Coale (1982) determined these values for α and β in the following manner. Stable populations of age 50 and over were created by combining four different positive annual growth rates with 32 Gompertzian life tables corresponding to mortality levels 9, 13, 17, and 21 for males and females of the South, East, and West Model Life Tables in Coale and Demeny (1966). Using these stable populations, the natural logarithm of the reciprocal of $e(a)M(a^+)$ was regressed on $r(a^+)[M(a^+)]^{-\alpha}$, with the condition that the intercept be zero, for each of a number of prefixed values of α.]

18. Mitra (1984) has derived a mathematical expression for the life expectancy at age a in terms of the death rate, the average age, and the rate of growth of the population in the age interval a

and above, assuming that the population aged a and above is stable. Mitra's formula is

$$e(a) = \frac{1}{M(a^+,r)} \exp\left[\frac{\{M(a^+,r) + r\}\{\bar{x}_s(a^+,r) - a\} - 1}{(1/r)M(a^+,r)}\right]$$

where $e(a)$ stands for the life expectancy at age a, $M(a^+, r)$ the age-specific death rate for the age group $[a, \infty)$ given stable growth rate r, and $\bar{x}_s(a^+, r)$ the average age of the population in the age group $[a, \infty)$. Using data from any of the *Demographic Yearbooks* published by the United Nations, estimate $e(65)$ according to Mitra's formula for a number of countries. Compare the estimates thus obtained with the corresponding estimates based on the Horiuchi–Coale approach (1982)

19. Suppose censuses are taken five years apart in a population. Consider censuses taken in years y and $y + 5$. Suppose we want to construct a life table applicable to the intercensus period, using the age data from these two censuses and the birth statistics during the intercensus years. We assume that the data are good. The following method is Coale's (1984). First we obtain by linear interpolation the age distribution for each intercensus year. By interpolating between the numbers reported at ages a in year y and $a + 5$ in year $y + 5$, we obtain intercensus estimates of the number of persons of age $a + 1$ in year $y + 1$, $a + 2$ in year $y + 2$, $a + 3$ in year $y + 3$, and $a + 4$ in year $y + 4$. The number of births reported for an intercensus year is taken as our estimate of the corresponding intercensus population of age 0. And interpolation of the number at age zero in year $y + 1$ and the census count at age 4 in year $y + 5$ gives the number of persons of age 1 in year $y + 2$, 2 in year $y + 3$, and 3 in year $y + 4$, and so on. Let $_1n_{x,y+k}$ be the number of persons of age $[x, x + 1)$ in year $y + k$, and let $*N(x)$ be the sum of $_1n_{x,y+k}$ over k (1 to 4) and the two census counts $_1n_{x,y}$ and $_1n_{x,y+5}$. The next step is to construct a preliminary life table using the formula

$$*l_{x+1}/*l_x = (*N(x + 1)/*N(x)) \exp(_1r_x)$$

with

$$_1r_x = \frac{_1n_{x,y+5} - _1n_{x,y}}{[*N(x) + *N(x + 1)]/2}$$

The third step is to use the preliminary life table thus constructed to revise the intercensus population estimates, assuming that deaths for each cohort during the intercensus period are distributed in accordance with the corresponding distribution of deaths in the life table population. These three steps are then repeated until the result converges.

Bibliographic Notes

A good understanding of the techniques of life table construction requires a firm grasp of the methods of estimation of the number exposed to risk. Actuarial textbooks cover this aspect in great detail (see, e.g., Gershenson, 1961; Batten, 1978). Familiarity with graduation techniques is also a necessary prerequisite (see, e.g., Miller, 1946). Chapter 15 of Shryock and Siegel (1973) provides a detailed exposition of the various steps involved in the construction of life tables. Another useful reference is Spiegelman (1968). Additional references may include Benjamin and Haycocks (1970), Chiang (1968, 1972), Cox (1975), and Jordan (1967).

Chapter 3 | The Mathematical Basis of the Life Table

The objective of this chapter is to highlight the mathematical basis of the life table. We shall treat x as a continuous rather than discrete variable and use the notation $l(x)$ in place of l_x. We denote by $h(x)$ the instantaneous death rate, also known as the *hazard rate*, or *force of mortality*. [A more common notation for the force of mortality is $\mu(x)$]. It is defined as the limit as Δx tends to zero of the ratio of the difference $l(x) - l(x + \Delta x)$ to the product $l(x) \Delta x$, i.e.,

$$h(x) = \lim_{\Delta x \to 0} \frac{l(x) - l(x + \Delta x)}{l(x) \Delta x} \tag{3.1}$$

or equivalently

$$h(x) = -l'(x)/l(x) \tag{3.2}$$

where the prime denotes the first derivative. Treating (3.2) as a differential equation, we obtain the solution

$$l(x) = l(0) \exp\left[-\int_0^x h(u) \, du \right] \tag{3.3}$$

where $l(0)$ is the radix of the life table. If $l(0) = 1$, we have from (3.3)

$$l(x) = \exp\left[-\int_0^x h(u) \, du \right] \tag{3.4}$$

We notice from (3.1) that $h(x)$ is also equal to the limit of the ratio $_{\Delta x} q_x / \Delta x$ as Δx tends to zero, where $_{\Delta x} q_x$ is the conditional probability of dying in the age interval x to $x + \Delta x$, given survival till age x. We may express (3.1) alternatively as

$$_{\Delta x} q_x = \Delta x \, h(x) + o(\Delta x) \tag{3.5}$$

where $o(\Delta x)$ is a function of Δx such that the ratio $o(\Delta x)/\Delta x$ tends to zero as Δx tends to zero. From (3.5) it follows that for very small values of Δx, the conditional probability of dying in the age interval x to $x + \Delta x$, given survival until x, is closely approximated by $\Delta x\, h(x)$.

1   Mortality Patterns

The age pattern of mortality can be described in terms of any of the three functions introduced above: $l(x)$, $h(x)$, or $_{\Delta x}q_x$. We shall discuss the following special patterns, with reference to the age interval from x to $x + 1$.

   i. $l(a)$ is linear in a,
  ii. the reciprocal of $l(a)$ is linear in a, and
 iii. the natural logarithm of $l(a)$ is linear in a.

All of these patterns are commonly discussed in actuarial textbooks (see, e.g., Batten, 1978). It is easy to verify that these patterns imply, respectively, an increasing, a decreasing, and a constant force of mortality. Thus, given $l(a) = A + Ba$, say, we have

$$h(a) = -l'(a)/l(a) = -B/l(a)$$

and, consequently,

$$h'(a) = [B/l(a)]^2 > 0$$

showing that $h(a)$ is an increasing function of a. Similarly, given that

$$[l(a)]^{-1} = F + Ga$$

we have

$$h'(a) = -[Gl(a)]^2 < 0$$

implying that $h(a)$ is a decreasing function, and given

$$l(a) = \exp[U + Va]$$

we have

$$h'(a) = 0$$

which means that $h(a)$ is a constant. Obviously, a force of mortality that remains constant or decreases continuously throughout the human lifespan is implausible. But there is nothing implausible about the assumption that the force of mortality is constant or decreasing in a small segment of the lifespan, irrespective of where that segment is. When modeling mortality patterns, one allows discontinuity, piecing together differing patterns assumed to prevail in different segments of the lifespan.

To pursue the comparative analysis of the three patterns a little further, let us consider the unit age interval mentioned above, namely, x to $x + 1$, where x is not necessarily an integer. No new principle is involved if we were to focus instead on a fraction of this unit age interval. A certain degree of notational simplicity can, however, be achieved by focusing on an interval of unit length. [We use the conventional notation q_x for $_1q_x$.]

Under pattern (i), it is easily shown that

$$_{1-t}q_{x+t} = (1 - t)q_x/[1 - tq_x] \tag{3.6}$$

Similarly, under pattern (ii), we have

$$_{1-t}q_{x+t} = (1 - t)q_x \tag{3.7}$$

and under pattern (iii)

$$_{1-t}q_{x+t} = 1 - \exp[-\mu(1 - t)] \tag{3.8}$$

where

$$\mu = -\ln[1 - q_x]$$

is the constant force of mortality characterizing the pattern.

Now imagine that A lives were under observation at exact age x, B lives entered the study at exact age $(x + r)$, C lives were censored (e.g., lost to follow up) at exact age $(x + s)$, and D lives terminated in death while under observation. Using this data let us estimate q_x under each of the three patterns.

Equating expected number of deaths to the observed number of deaths we obtain the basic estimating equation

$$Aq_x + B_{1-r}q_{x+r} - C_{1-s}q_{x+s} = D \tag{3.9}$$

which is specialized under pattern (i) to

$$Aq_x + \frac{B(1 - r)q_x}{1 - rq_x} - \frac{C(1 - s)q_x}{1 - sq_x} = D \tag{3.10}$$

under pattern (ii) to

$$Aq_x + B(1 - r)q_x - C(1 - s)q_x = D \tag{3.11}$$

and under pattern (iii) to

$$A(1 - e^{-\mu}) + B[-e^{-(1-r)\mu}] + C[1 - e^{-(1-s)\mu}] = D \tag{3.12}$$

where $\mu = -\ln[1 - q_x]$. Equation (3.11) gives the closed-form solution for q_x

$$q_x = D/[A + B(1 - r) - C(1 - s)] \tag{3.13}$$

To solve Eqs. (3.10) and (3.12) we have to use iterative methods, since closed-form solutions are not available. Let us consider a numerical example.

Suppose

$$A = 1{,}000; \qquad B = 100; \qquad \text{and} \qquad C = 50;$$

$$D = 12; \qquad r = 0.25; \qquad \text{and} \qquad s = 0.75$$

From (3.13) we notice that, under pattern (ii), our estimate of q_x is

$$12/[1{,}000 + 100 \times 0.75 - 50 \times 0.25] = 0.0112941,$$

whereas the corresponding estimates obtained by iterative methods from (3.10) and (3.12) are 0.0112877 and 0.0112936, respectively. These three estimates are impressively close to each other.

Obviously, ease of calculation favors pattern (ii). A second reason why some analysts prefer pattern (ii) is that the structure of the numerator and denominator on the right-hand side of (3.13) can be readily interpreted. Let us regard observing human lives during an age interval of unit length as analogous to subjecting an electric appliance to one cycle of use in a life-testing experiment. It is then easily seen that the denominator of the right-hand side of (3.13) is analogous to the total number of cycles of appliance use in the experiment, while the numerator is analogous to the number of cycles that terminate in appliance failure. [Notice that if there were no censoring the denominator on the right-hand side of (3.13) would have been

$$A q_x + B_{1-r} q_{x+r}$$

The censoring necessitates the subtraction of $C_{1-s} q_{x+s}$ from this number in order for the denominator to correspond to what may be called the effective number of use cycles in the life testing experiment.]

In Table 3.1 some of the characteristic features of the three patterns are shown.

Table 3.1

Comparative Properties of Three "Popular" Mortality Patterns in the Age Interval x to $x + 1$

	Pattern i: $l(a)$ is linear	Pattern ii[a]: $1/l(a)$ is linear	Pattern iii: natural log of $l(a)$ is linear
$h(x + a)$	$\dfrac{q_x}{1 - aq_x}$	$\dfrac{q_x}{1 - (1 - a)q_x}$	Invariant
$_{1-a}q_{x+a}$	$\dfrac{(1 - a)q_x}{1 - aq_x}$	$(1 - a)q_x$	$1 - e^{-(1-a)\mu}$

[a] Pattern ii was suggested by the Italian actuary Gaetano Balducci in 1920, and is often referred to in the actuarial literature as the Balducci hypothesis.

The expression for $h(a + x)$ under pattern (ii) deserves a comment. In the context of the example given earlier (A lives under observation at exact age x, B lives entering observation at exact age $x + r$, C lives getting censored at exact age $x + s$, and D lives terminating in death during observation) recall that our estimate of q_x was D/E, where $E = A + (1 - r)B - (1 - s)C$. Substitution gives (see Table 3.1)

$$h(x + a) = \frac{D}{A + (1 - r)B - (1 - s)C - (1 - a)D}$$

Note that the denominator on the right-hand side of this equation can be obtained by placing D at exact age $x + a$ and then treating them exactly as though they were censored cases.

2   The Abridged Life Table

In this section we shall present a formula derived by Keyfitz and Frauenthal (1975) for $l(x + n)/l(x)$ in terms of observed data and show how other formulas proposed earlier by other scholars follow from it.

We start with the definition of age-specific death rate for the age group $[x, x + n]$

$$_nM_x = \int_x^{x+n} p(a)h(a)\, da \bigg/ \int_x^{x+n} p(a)\, da \tag{3.14}$$

where $p(a)$ represents population of age a, and $h(a)$ the force of mortality at age a, both pertaining to a given time point. If $p(a)$ is proportional to the life table suvivor function and if the latter is linear, then it is easily verified that (3.14) simplifies to

$$_nM_x = \frac{l(x) - l(x + n)}{n[l(x) + l(x + n)]/2} \tag{3.15}$$

[In deriving (3.15) we use the definition $l(x + a)h(x + a) = -l'(x + a)$, and the knowledge that if $l(a)$ is linear, the area under $l(a)$ between x and $x + n$ is equal to n times the ordinate of $l(a)$ at the mid-point between x and $x+n$.] From (3.15) it follows that

$$\frac{l(x + n)}{l(x)} = \frac{1 - (n/2)_nM_x}{1 + (n/2)_nM_x} \tag{3.16}$$

giving

$$_nq_x = 1 - \frac{l(x + n)}{l(x)} = \frac{_nM_x}{1 + (n/2)_nM_x} \tag{3.17}$$

To obtain the corresponding results under a somewhat more general setup, let us assume that both $p(a)$, the population-at-risk function, and $h(a)$, the hazard function (force of mortality), are linear in the age interval $x - n$ to $x + 2n$. Note that this is not an untenable assumption, since there is no necessary relationship between $p(a)$ and $h(a)$. Under the linearity assumption, we may replace $p(a)$ and $h(a)$ in the age interval x to $x + n$ by

$$p(x + 0.5n) + p'(x + 0.5n)(a - x - 0.5n) \tag{3.18}$$

and

$$h(x + 0.5n) + h'(x + 0.5n)(a - x - 0.5n) \tag{3.19}$$

respectively. Moreover, because of the linearity assumption

$$p'(x - 0.5n) = [p(x + 1.5n) - p(x - 0.5n)]/2n \tag{3.20}$$

But linearity implies also that $np(x + 1.5n) = {}_nP_{x+n}$ and $np(x - 0.5n) = {}_nP_{x-n}$, where

$$_nP_{x+t} = \int_{x+t}^{x+t+n} p(a)\, da \tag{3.21}$$

Hence, from (3.20)

$$p'(x + 0.5n) = ({}_nP_{x+n} - {}_nP_{x-n})/2n^2 \tag{3.22}$$

Similarly,

$$h'(x + 0.5n) = [h(x + 1.5n) - h(x - 0.5n)]/2n \tag{3.23}$$

which leads to, on replacement of $h(x + 1.5n)$ by ${}_nM_{x+n}$ and $h(x - 0.5n)$ by ${}_nM_{x-n}$,

$$h'(x + 0.5n) = ({}_nM_{x+n} - {}_nM_{x-n})/2n \tag{3.24}$$

It is now easy to verify that

$$\int_x^{x+n} p(a)h(a)\, da = np(x + 0.5n)h(x + 0.5n)$$
$$+ (n^3/12)p'(x + 0.5n)h'(x + 0.5n) \tag{3.25}$$

If on the right-hand side of (3.25), we replace $np(x + 0.5n)$ by ${}_nP_x$ and $h(x + 0.5n)$ by $(1/n)$ of the integral of $h(a)$ from x to $x + n$ and then use the definition of ${}_nM_x$ and equations (3.22) and (3.24), we obtain

$$_nM_x = \left(\frac{1}{n}\right)\int_x^{x+n} h(a)\, da + \frac{1}{48\,{}_nP_x}({}_nP_{x+n} - {}_nP_{x-n})({}_nM_{x+n} - {}_nM_{x-n}) \tag{3.26}$$

which gives recalling that $h(x) = -l'(x)/l(x)$

$$l(x + n)/l(x) = \exp[-n(_nM_x + C)] \tag{3.27}$$

where

$$C = (n/48 \, _nP_x)(_nP_{x-n} - \, _nP_{x+n})(_nM_{x+n} - \, _nM_{x-n}) \tag{3.28}$$

This result is due to Keyfitz and Frauenthal (1975). The utility of this result can be appreciated if we notice that Greville's (1943) well-known expression for $l(x + n)/l(x)$ can be derived from (3.25) in the following manner: In (3.25) replace $P(x)$ with $l(x)$; then the integral on the left-hand side divided by the corresponding integral of $l(x)$ is m_x, the life table death rate. On the right-hand side of (3.25) replace $l'(x)^n h'(x)$ with $-l(x)[h(x)]^2[h'(x)/h(x)]$, remembering that $l'(x) = -l(x)h(x)$. Also approximate $l(x + 0.5n)/_nL_x$ with $1/n$ and $h(x + 0.5n)$ with $_nm_x$. These steps lead to the approximation

$$n_nm_x + (n^3/12)(_nm_x)^2(\ln \, _nm_x)' \tag{3.29}$$

for the area under the force of mortality function from x to $x + n$, from which we obtain $l(x + n)/l(x)$ in the usual manner. If we now use the empirical regularity that for most of the age groups in most life tables $(\ln \, _nM_x)'$ remains constant at about -0.096, we get from the result just obtained the approximation for $l(x + n)/l(x)$

$$\frac{l(x + n)}{l(x)} = \exp\left[-\int_0^n h(x + a) \, da\right] = \exp[-n_nm_x - 0.008n^3 \, _nm_x^2] \tag{3.30}$$

which is the widely used Reed–Merrell (1939) formula.

Expected Fraction of the Last n Years of Life

The relationship between $_nq_x$ and $_nm_x$ of the life table can be expressed in terms of what may be called the expected fraction of the last n years of life lived. Recall that $_nL_x$ is equal to the total number of years of life lived by $l(x)$ individuals over the age interval x to $x + n$. This, it may be noted, consists of two parts: (1) n full years lived by each of $l(x + n)$ survivors till age $x + n$ out of $l(x)$, and (2) n times the fraction of n years lived, on average, by each of those dying in the age interval to x to $x + n$. That is, using the notation nf_x for the fraction just mentioned,

$$_nL_x = nl(x + n) + n_nf_{x\,n}d_x$$
$$= n[l(x) - l(x)_nq_x] + n_nf_xl(x)_nq_x$$
$$= nl(x)[1 - (1 - \, _nf_x)_nq_x] \tag{3.31}$$

whence

$$_nm_x = \, _nd_x/_nL_x = l(x)_nq_x/_nL_x = \, _nq_x/n[1 - (1 - \, _nf_x)_nq_x] \tag{3.32}$$

or

$$_nq_x = {}_nn_mx_x/[1 + n(1 - {}_nf_x)_nm_x]$$ (3.33)

(see Chiang, 1960a). If one knows the values of the fractions $_nf_x$, the relationship (3.33) can be used for estimating $_nq_x$ by replacing $_nm_x$ by the corresponding observed age-specific death rate. Chiang has assembled some data indicating that the sequence of fractions $_nf_x$ shows little variation over subpopulations defined in terms of race, region of residence, etc. Not all scholars are convinced that the sequence has the invariance property attributed to it.

3   Some Mathematical Relationships among the Life Table Functions

We have already seen that the definition of the force of mortality $[h(x) = l'(x)/l(x)]$ implies

$$l(x) = l(0) \exp\left[\int_0^x h(a)\, da\right]$$ (3.34)

Now with $l(0) = 1$

$$f(x) = h(x)l(x)$$

is the probability density function of the age at death. Note that

$$\int_0^\infty f(a)\, da = \int_0^\infty h(a)l(a)\, da = -l(a)\Big|_0^\infty = 1$$ (3.35)

and that

$$\int_x^{x+n} f(a)\, da = -\int_x^{x+n} l'(a)\, da = -l(a)\Big|_x^{x+n} = l(x) - l(x + n) = {}_nd_x$$ (3.36)

We thus have, with $l(0) = 1$,

$$l(x) = \exp\left[-\int_0^x h(a)\, da\right]$$ (3.37)

giving $l(x)$ in terms of past lifetime, and

$$l(x) = \text{Pr (the age at death is } x \text{ or higher)}$$

$$= \int_x^\infty f(a)\, da = \int_x^\infty h(a)l(a)\, da$$

giving $l(x)$ in terms of future lifetimes.

The conditional probability of dying in the age interval $[x, x + n)$, given survival until age x (i.e., $_nq_x$), is the same as the probability that the age at death is in $[x, x + n)$, conditional on it being x or higher. We thus have

$$_nq_x = \frac{\int_x^{x+n} f(a)\, da}{\int_x^\infty f(a)\, da} = \frac{l(x) - l(x + n)}{l(x)} \tag{3.39}$$

It follows that the conditional probability of surviving till age $x + n$, given survival until age x is

$$_np_x = 1 - {}_nq_x = \frac{l(x + n)}{l(x)} = \frac{\exp[-\int_0^{x+n} h(a)\, da]}{\exp[-\int_0^x h(a)\, da]} \tag{3.40}$$

and for a set of points $0 < s < t < u < v < w < x$, if $l(0) = 1$, we have

$$l(x) = \exp\left[-\int_0^x h(a)\, da\right]$$

$$= \exp\left[\int_0^s h(a)\, da\right] \exp\left[-\int_s^t h(a)\, da\right] \cdots \exp\left[-\int_w^x h(a)\, da\right] \tag{3.41}$$

Now, denoting the age at death by the random variable X, its expected value, conditional on dying after attaining age x, is given by

$$E(X \mid X \geq x) = \frac{\int_x^\infty a f(a)\, da}{\int_x^\infty f(a)\, da} = \frac{\int_x^\infty a[-l'(a)]\, da}{\int_x^\infty [-l'(a)]\, da} = x + \frac{T(x)}{l(x)} \tag{3.42}$$

where

$$T(x) = \int_x^\infty l(a)\, da \tag{3.43}$$

Noting that the expected value of the lifetime remaining after attaining age x is equal to the expected value of the age at death, conditional on dying after attaining age x, minus x, we obtain from (3.42)

$$\text{life expectancy at age } x = T(x)/l(x) \tag{3.44}$$

Similarly, the expected value of the age at death, conditional on dying in the age interval $[x, x + n)$, is

$$E(X \mid x \leq X < x + n) = \frac{\int_x^{x+n} a f(a)\, da}{\int_x^{x+n} f(a)\, da} = x + \frac{_nL_x - n l(x + n)}{_nd_x} \tag{3.45}$$

$$= x + {}_na_x$$

where

$$_nL_x = \int_x^{x+n} l(a)\, da, \quad _nd_x = l(x) - l(x + n), \quad \text{and} \quad _na_x = \frac{_nL_x - n l(x + n)}{_nd_x}$$

This last component represents the expected length of the lifetime lived in the age interval $[x, x + n)$, conditional on dying in that age interval. Note that the fraction $_nf_x$ introduced earlier and $_na_x$ just defined are related to each other in the following manner:

$$_na_x = n_nf_x \qquad (3.46)$$

Starting from the definition of $_na_x$ given above, we can derive a formula for decomposing the expectation of life at age x into two parts. Notice from the definition of $_na_x$ that

$$_nL_x = {}_na_{x\,n}d_x + nl(x + n)$$

and recall that

$$T(x) = {}_nL_x + T(x + n)$$

Substitution of the latter into the former gives

$$T(x) = {}_na_{x\,n}d_x + nl(x + n) + T(x + n)$$

from which, if we divide throughout by $l(x)$, we obtain

life expectancy at age x

$$= {}_na_{xn}q_x + (n + \text{life expectancy at age } x + n)_np_x \qquad (3.47)$$

which decomposes the expectation of life at age x into two parts pertaining, respectively, to the lifetime in the age interval $[x, x + n)$ and that in the age interval $[x + n, \infty)$. One could use this result to decompose into meaningful parts the difference between the expectations of life at age x from two life tables (see, e.g., Keyfitz, 1968).

4   The Life Table from the Perspective of Markov Processes

In Chapter 9, the multistate life table is discussed using the framework of continuous time Markov processes. To prepare the reader for that discussion, we present in this section a view of the ordinary life (mortality) table from the perspective of continuous time Markov processes.

Let $X(t)$ be a random variable that takes only discrete values to which are attached positive integers for easy reference. In the context of the ordinary life table, $X(t)$ takes only two values: 1, alive (or surviving); and 2, dead. Note that these are exhaustive and mutually exclusive possibilities. The collection of (exhaustive and mutually exclusive) discrete values that $X(t)$ takes is referred to as the *state space*, and the elements of the state space are called *states*. The time parameter t in $X(t)$ is continuous. But it need not necessarily refer to calendar time. For example, in the ordinary life table, it refers to age.

When describing $X(t)$ in its dynamic aspects, phraseology such as the following is often used: "the process occupies state i at time t," "if state k is entered at any time, the process remains there indefinitely," and so on. The term process in such phrases may be interpreted to mean, in the context of the ordinary life table, an individual randomly selected from those who are subjected to the mortality pattern represented by the life table.

Very often the description of the dynamic features of $X(t)$ centers around comparisons of the states occupied by the process at two points in time. For this purpose, we introduce the notation $q_{ij}(s, t)$ to denote the conditional probability that the process occupies state j at time t, given that it was occupying state i at time s, where $0 \le s \le t$. These probabilities are known as *transition probabilities*.

It is useful to arrange the transition probabilities in a row-by-column fashion, i.e., in the form of a matrix. For the ordinary life table the arrangement looks like the following:

	State occupied at time t	
State occupied at time s ($\le t$)	1 (alive)	2 (dead)
1 (alive)	$q_{11}(s, t)$	$q_{12}(s, t)$
2 (dead)	$q_{21}(s, t)$	$q_{22}(s, t)$

Obviously here $q_{21}(s, t) = 0$, because transition from the state "dead" to the state "alive" is impossible. For the same reason, $q_{22}(s, t) = 1$. (States from which escape is not possible are called *absorbing states*, as contrasted with others that are *transient*.)

Note that each row of the matrix of transition probabilities sum to unity. This is simply a consequence of the fact that the entries in each row are conditional probabilities, pertaining to exhaustive and mutually exclusive possibilities.

We shall refer to the two-by-two arrangement shown above by the symbol $\mathbf{Q}(s, t)$:

$$\mathbf{Q}(s, t) = \begin{bmatrix} q_{11}(s, t) & q_{12}(s, t) \\ q_{21}(s, t) & q_{22}(s, t) \end{bmatrix} = \begin{bmatrix} q_{11}(s, t) & q_{12}(s, t) \\ 0 & 1 \end{bmatrix}$$

Note that $q_{12}(s, t)$ herein corresponds to $_{t-s}q_s$ and $q_{11}(s, t)$ to $_{t-s}p_s$ or $1 - {}_{t-s}q_s$, in the conventional notation introduced earlier.

The mortality process can be described also in terms of what are known as *transition intensities*. We define (for i not equal to j)

$$r_{ij}(s) = \lim_{u \to 0} q_{ij}(s, s + u)/u \qquad (3.48)$$

to represent the intensity (force) of transition from state i to state j at time s. We introduce for convenience $r_{ii}(s)$ such that $\Sigma \, r_{ij}(s) = 0$, where the summation is over j. It is easy to verify that, thus defined, $r_{ii}(s) = \lim\{[q_{ii}(s, s + u) - 1]/u\}$ as $u \to 0$. If we set the transition intensities in a matrix form so as to correspond to $\mathbf{Q}(s, t)$ introduced above we have

$$\mathbf{R}(s) = \begin{bmatrix} r_{11}(s) & r_{12}(s) \\ r_{21}(s) & r_{22}(s) \end{bmatrix}$$

Clearly each row of $\mathbf{R}$ sums to zero. Also, in the present case, $r_{21}(s) = r_{22}(s) = 0$, because $q_{21}(s, t) = 0$ and $q_{22}(s, t) = 1$.

Let us now consider three time points $s \leq t \leq t + u$. The transitions during the first and last segments can be described in terms of the following transition probability matrices:

$$\mathbf{Q}(s, t) = \begin{bmatrix} q_{11}(s, t) & q_{12}(s, t) \\ 0 & 1 \end{bmatrix}$$

and

$$\mathbf{Q}(t, t + u) = \begin{bmatrix} q_{11}(t, t + u) & q_{12}(t, t + u) \\ 0 & 1 \end{bmatrix}$$

whereas those during the whole interval s to $t + u$ can be described in terms of

$$\mathbf{Q}(s, t + u) = \begin{bmatrix} q_{11}(s, t + u) & q_{12}(s, t + u) \\ 0 & 1 \end{bmatrix}$$

An important relationship exists between these three transition matrices:

$$\mathbf{Q}(s, t + u) = \mathbf{Q}(s, t)\mathbf{Q}(t, t + u) \tag{3.49}$$

To verify this we examine whether the elements of the product on the right-hand side are equal to the corresponding elements on the left-hand side of the equation. The product on the right-hand side of (3.49) is

$$\begin{bmatrix} q_{11}(s, t)q_{11}(t, t + u) & q_{11}(s, t)q_{12}(t, t + u) + q_{12}(s, t) \\ 0 & 1 \end{bmatrix}$$

Therefore the relationship (3.49) holds if and only if both of the following relationships hold:

$$q_{11}(s, t + u) = q_{11}(s, t)q_{11}(t, t + u) \tag{3.50}$$

$$q_{12}(s, t + u) = q_{11}(s, t)q_{12}(t, t + u) + q_{12}(s, t) \tag{3.51}$$

That these latter relationships hold follows from probability laws. For (3.50) simply says that the probability of surviving to age $t + u$, given survival to age s, is equal to the product of the following two probabilities: that of surviving to age t, given survival to age s; and that of surviving to age $t + u$, given survival to age t. And (3.51) simply says that the probability of dying after attaining age s but before reaching age $t + u$ is equal to the probability of dying in the age interval s to t, given survival to age s, *plus* the product of the following two probabilities: that of surviving from age s to t, given survival to age s, and that of dying between ages t and $t + u$, given survival to age t.

From (3.50) to (3.51) we can derive what are known as the Chapman–Kolmogorov forward equations. Subtracting $q_{11}(s, t)$ from both sides of (3.50) and then dividing by u gives

$$(q_{11}(s, t + u) - q_{11}(s, t))/u = q_{11}(s, t)(q_{11}(t, t + u) - 1)/u \tag{3.52}$$

Now taking the limit as u tends to zero gives

$$\partial q_{11}(s, t)/\partial t = q_{11}(s, t)r_{11}(t) \tag{3.53}$$

Similarly from (3.51) [by subtracting $q_{12}(s, t)$ from both sides, then dividing by u, and finally taking limit as u tends to zero] we get

$$\partial q_{12}(s, t)/\partial t = q_{11}(s, t)r_{12}(t) \tag{3.54}$$

Recalling that $q_{21}(s, t) = 0$, $q_{22}(s, t) = 1$, and $r_{21}(t) = r_{22}(t) = 0$, we obtain the following matrix equation, given (3.53) and (3.54):

$$\begin{bmatrix} \partial q_{11}(s, t)/\partial t & \partial q_{12}(s, t)/\partial t \\ 0 & 0 \end{bmatrix} = \begin{bmatrix} q_{11}(s, t) & q_{12}(s, t) \\ 0 & 1 \end{bmatrix} \begin{bmatrix} r_{11}(t) & r_{12}(t) \\ 0 & 0 \end{bmatrix}$$

We may expresss this matrix equation alternatively as

$$\partial \mathbf{Q}(s, t)/\partial t = \mathbf{Q}(s, t)\mathbf{R}(t) \tag{3.55}$$

where taking a partial derivative of a matrix is understood to involve taking partial derivatives of its elements.

If we now consider the three time (age) points $s \le s + v \le t$ and proceed in the manner outlined above, we get

$$\mathbf{Q}(s, t) = \mathbf{Q}(s, s + v)\mathbf{Q}(s + v, t) \tag{3.56}$$

or, equivalently,

$$q_{11}(s, t) = q_{11}(s, s + v)q_{11}(s + v, t) \tag{3.57}$$

and

$$q_{12}(s, t) = q_{11}(s, s + v)q_{12}(s + v, t) + q_{12}(s, s + v) \tag{3.58}$$

By subtracting $q_{11}(s + v, t)$ from both sides of (3.58), then dividing both sides by v, and finally taking limits as v tends to zero, we obtain

$$\partial q_{11}(s, t)/\partial s = -q_{11}(s, t)r_{11}(s) \tag{3.59}$$

and, similarly, by subtracting $q_{12}(s + v, t)$ from both sides of (3.59), then dividing by v, and finally taking limits as v tends to zero, we obtain

$$\partial q_{12}(s, t)/\partial s = -q_{12}(s, t)r_{11}(s) - r_{12}(s) \tag{3.60}$$

We thus have the following matrix equation:

$$\begin{bmatrix} \partial q_{11}(s, t)/\partial s & \partial q_{12}(s, t)/\partial s \\ 0 & 0 \end{bmatrix} = -\begin{bmatrix} r_{11}(s) & r_{12}(s) \\ 0 & 0 \end{bmatrix}\begin{bmatrix} q_{11}(s, t) & q_{12}(s, t) \\ 0 & 1 \end{bmatrix}$$

or, equivalently,

$$\partial \mathbf{Q}(s, t)/\partial s = -\mathbf{R}(s)\mathbf{Q}(s, t) \tag{3.61}$$

This is known as the Chapman–Kolmogorov backward system.

Before discussing the implications of the two Chapman–Kolmogorov systems, let us satisfy ourselves that the force of mortality $h(t)$ introduced earlier in the chapter is the same as the transition intensity $r_{12}(t)$ introduced in this section. Recall that, in the context of the ordinary life table, the unconditional probability that the process is alive at age t is given by $l(t)$, provided $l(0) = 1$. We also know that for $s \leq t$

$$l(t) = l(s)q_{11}(s, t) \tag{3.62}$$

If we differentiate both sides of (3.62) with respect to t, we get

$$l'(t) = l(s)\, \partial q_{11}(s, t)/\partial t \tag{3.63}$$

Substitution for the second factor on the right-hand side of (3.63) from (3.53) gives $l'(t) = l(s)q_{11}(s, t)r_{11}(t)$, which is equivalent to $l'(t) = l(t)r_{11}(t)$, by virtue of (3.62). But $r_{11}(t) = -r_{12}(t)$ because the row sums of the $\mathbf{R}$ matrix are all zero. We thus have

$$l'(t) = -l(t)r_{12}(t) \tag{3.64}$$

demonstrating that $h(t) = -l'(t)/l(t)$, introduced earlier, is the same as $r_{12}(t)$. The complement of $l(t)$ is $1 - l(t)$, the unconditional probability that the process is in the dead state as of age (time) t. Corresponding to (3.64) we have

$$d[1 - l(t)]/dt = l(t)r_{12}(t) \tag{3.65}$$

Stacking (3.64) and (3.65) together in matrix form gives

$$\frac{d}{dt}\begin{bmatrix} l(t) & 1 - l(t) \end{bmatrix} = \begin{bmatrix} l(t) & 1 - l(t) \end{bmatrix}\begin{bmatrix} r_{11}(t) & r_{12}(t) \\ 0 & 0 \end{bmatrix}$$

which is in a form generalizable to the multistate context.

We now return to the Chapman–Kolmogorov equations to derive from them estimating equations for the transition probabilities and the transition intensities. From (3.53) we obtain on integration

$$q_{11}(s, t) = 1 + \int_s^t q_{11}(s, x)r_{11}(x)\, dx \qquad (3.66)$$

using the initial condition that $q_{11}(t, t) = 1$. Similarly from (3.54) we obtain

$$q_{12}(s, t) = \int_s^t q_{11}(s, x)r_{12}(x)\, dx \qquad (3.67)$$

We thus have the following matrix equation

$$\begin{bmatrix} q_{11}(s, t) & q_{12}(s, t) \\ 0 & 1 \end{bmatrix} = \begin{bmatrix} 1 & 0 \\ 0 & 1 \end{bmatrix} + \begin{bmatrix} \int_s^t q_{11}(s, x)r_{11}(x)\, dx & \int_s^t q_{11}(s, x)r_{12}(x)\, dx \\ 0 & 0 \end{bmatrix}$$

or

$$\mathbf{Q}(s, t) = \mathbf{I} + \int_s^t \mathbf{Q}(s, x)\mathbf{R}(x)\, dx \qquad (3.68)$$

Similarly from (3.59) we get

$$q_{11}(s, t) = 1 - \int_s^t q_{11}(x, t)r_{11}(x)\, dx \qquad (3.69)$$

and from (3.60)

$$q_{12}(s, t) = - \int_s^t [q_{12}(x, t)r_{11}(x) + r_{12}(x)]\, dx \qquad (3.70)$$

or, putting these together in a matrix equation,

$$\mathbf{Q}(s, t) = \mathbf{I} + \int_s^t \mathbf{R}(x)\mathbf{Q}(x, t)\, dx \qquad (3.71)$$

For estimating the transition intensities, we segment the age (time) dimension into a number of intervals, each of which is small enough to justify the assumption that within it, the transition intensity is constant. If $[s, s + u)$ is one such interval, we have from (3.67)

$$q_{12}(s, s + u) = r_{12}^*(s, s + u) \int_s^{s+u} q_{11}(s, x)\, dx \qquad (3.72)$$

where we assume that $r_{12}(x) = r_{12}^*(s, s + u)$ for x in the interval s to $s + u$. From (3.72) we obtain by division the following estimating formula

$$r_{12}^*(s, s + u) = q_{12}(s, s + u) \bigg/ \int_s^{s+u} q_{11}(s, x)\, dx \qquad (3.73)$$

the right-hand side of which has the structure of the usual occurrence–exposure ratio. This suggests that the usual central age-specific death rate in the interval $[s, s + u)$ can be taken as an estimate of $r^*_{12}(s, s + u)$. Once $r^*_{12}(s, s + u)$ has been estimated, we estimate $r^*_{11}(s, s + u)$ from the relationship $r^*_{11}(,) + r^*_{12}(,) = 0$.

For the interval $[s, s + u)$, in which the intensity rate is assumed to remain constant, we have on integration, from (3.53),

$$q_{11}(s, s + u) = \exp[r^*_{11}(s, s + u)] \tag{3.74}$$

and from (3.54), using (3.53),

$$q_{12}(s, s + u) = [r^*_{12}(s, s + u)/r^*_{11}(s, s + u)][q_{11}(s, s + u) - 1] \tag{3.75}$$

These equations thus yield estimates of the elements of $Q(s, s + u)$. To estimate $Q(s, t)$ we partition $[s, t)$ into subintervals: $[s, t_1), [t_1, t_2), \ldots, [t_{n-1}, t)$, such that in each subinterval we may assume that the intensity of transition remains constant, thus permitting the application of (3.74) and (3.75). We then apply the relationship

$$Q(s, t) = Q(s, t_1)Q(t_1, t_2) \cdots Q(t_{n-1}, t) \tag{3.76}$$

We now turn to summary measures that one calculates from the ordinary life table. An often-used summary measure is the life expectancy at age x, which, it may be noted, is the expected duration the process stays alive beyond age x, given survival to (occupancy of the alive state at) age x. One could also consider the expected duration the process stays alive during the n years following age x, given survival to age x. Let us introduce the indicator random variable $I_{11}(s, t)$, which takes the value 1 if the process is alive (i.e., it occupies state 1) at age t, given that it was occupying state 1 at age s. Let it take the value 0 otherwise. Clearly, the expected duration the process remains alive in the interval s to $s + u$, is the expected value of

$$\int_s^{s+u} I_{11}(s, x) \, dx$$

which is

$$\int_s^{s+u} q_{11}(s, x) \, dx$$

5 Heterogeneity in Mortality Experiences of Individuals

Common sense tells us that individuals vary in their capacity to survive. The conventional approach to life table analysis recognizes this when separate life tables are constructed for males and females, blacks and whites, and so on.

In Chapters 11–13 we discuss procedures to take into account known risk factors that cause heterogeneity in mortality patterns among individuals in a population. In the present section we discuss a procedure to incorporate heterogeneity in the age pattern of mortality of individuals that cannot be attributed to a known risk factor (among the works dealing with this topic are Hougaard, 1984a,b; Keyfitz, 1984; Keyfitz and Littman, 1980; Manton and Stallard, 1981a,b; Manton *et al.*, 1981a; Nour, 1984; Vaupel *et al.*, 1979; Vaupel and Yashin, 1983, 1985).

A Simple Model

Assume that the force of mortality (hazard function) can be written as

$$h(t) = Zh_0(t) \tag{3.77}$$

where Z is a random variable that takes only positive values and $h_0(t)$ is assumed to be independent of Z. [The random variable Z may be thought of as representing human frailty that operates in a multiplicative manner on the force of mortality, the latter being common for the members of a population. The specification of $h(t)$ in (3.77) implies that conditional on $Z = z$, the survival function, is

$$S(x \mid Z = z) = \exp\left[-z \int_0^x h_0(t)\, dt \right] \tag{3.78}$$

or in terms of the cumulative hazard function $H_0(x)$ corresponding to $h_0(x)$,

$$S(x \mid Z = z) = \exp[-zH_0(x)]$$

Now suppose Z is distributed in the population with a probability density function (pdf) $g(z)$. Then, the unconditional survival function is given by

$$
\begin{aligned}
S(x) &= \int_0^\infty S(x \mid Z = z) g(z)\, dz \\
&= \int_0^\infty g(z) \exp[-zH_0(x)]\, dz
\end{aligned}
\tag{3.79}
$$

In particular, if $g(z)$ stands for the gamma density, i.e., if

$$g(z) = [\beta^\alpha / \Gamma(\alpha)] z^{\alpha-1} \exp(-\beta z) \tag{3.80}$$

then, as Vaupel *et al.* (1979) have shown,

$$S(x) = \beta^\alpha [\beta + H_0(x)]^{-\alpha} \tag{3.81}$$

The conditional distribution of Z, given survival until age x, can be derived in the general setup in the usual fashion noting that the joint distribution of X

and Z is given by the pdf

$$f(x, z) = zh_0(x)g(z) \exp[-zH_0(x)] \qquad (3.82)$$

It is easy to see that

$$g(z \mid X > x) = [S(x)]^{-1} \int_0^\infty f(x, z)\, dx \qquad (3.83)$$

$$= [S(x)]^{-1} g(z) \exp[-zH_0(x)]$$

Problems and Complements

1. If $q_x = 0.25$, find $_{0.5}q_{x+0.5}$ under each of the patterns in Table 3.1.

2. Express $_{1-s}q_{x+t}$ in terms of q_x under patterns (i) and (ii) and in terms of μ under pattern (iii) of Table 3.1. (See Batten, 1978, chapter 1.)

3. If $q_x = 0.25$, find $_{0.3}q_{x+0.5}$ under each of the patterns in Table 1.

(4). If pattern (ii) of Table 3.1 applies to a population, and $_{0.3}q_{x+0.5} = 0.02$, find q_x.

5. If $q_{30} = 0.1$, find the probability that a person born on February 1, of year y and surviving until April 1 of year $(y + 30)$ will die sometime between July 1 and December 31 of year $(y + 30)$, assuming pattern (ii) of Table 3.1.

6. If $l_x = 25,500$ and $l_{x+1} = 22,000$, how many deaths in the life table population would you expect between ages $x + 0.5$ and $x + 1$, given that $_tq_x = tq_x$, for $0 \le t \le 1$?

7. Given a complete life table, how would you estimate the probability that a person of exact age $x + \theta$ will die before age $x + n + \varphi$, where x and n are integers? [ANSWER: Estimate $_{n+(\varphi-\theta)}q_{x+\theta} = 1 - (l_{x+n+\varphi}/l_{x+\theta})$. One has to use interpolation to obtain values of $l_{x+n+\varphi}$ and $l_{x+\theta}$. Show how to interpolate under each of the patterns of Table 3.1.]

8. Let y denote the age up to which 50 percent of persons all of whom are of age x at present will survive. [That is, $y - x$ is the *median* remaining lifetime for the members of the group.] The value of y must satisfy the equation $l_y = l_x/2$. Given a complete life table, we can find an integer a such that $l_a < l_x/2 < l_{a+1}$. Linear interpolation gives $a + \theta$ between a and $a + 1$ such that $l_{a+\theta} = l_x/2$. The appropriate formula for interpolation depends upon the assumption one makes regarding the pattern of $l(x)$.

9. Given that Z is a random variable representing human frailty acting multiplicatively on the force of mortality $h_0(x)$ common for all members of a population, and that the joint probability density of X and Z is given by $f(x, z) = zg(z)h_0(x) \exp[-zH_0(x)]$, $g(z)$ being the pdf of Z, it follows that the conditional distribution of z, given survival till age x, is given by

$$g(z \mid X > x) = [S(x)]^{-1} \int_x^\infty f(x, z)\, dx$$

$$= [S(x)]^{-1} g(z) \exp[-zH_0(x)]$$

The mean value of Z conditional on $X > x$, is given by

$$E(Z \mid X > x) = [S(x)]^{-1} \int zg(z) \exp[-zH_0(x)]\, dz$$

10. Show that under the setup just described,

$$E(Z \mid X > x) = -[h_0(x)]^{-1} \, d \ln S(x)/dx$$

[HINT: $S(x) = \int g(z) \exp[-zH_0(x)] \, dz$, and $dH_0(x)/dx = h_0(x)$.]

11. Show that, if Z in Exercise 9 has the gamma density (3.80), then $E(z \mid X > x) = \alpha[\beta + H_0(x)]^{-1}$. [HINT: See (3.81) for $S(x)$, given that Z has a gamma density.]

12. Show that $E(z \mid X > x)$ is a decreasing function of x. [HINT: Show that the partial derivative of $E(z \mid X > x)$ with respect to x can be expressed as $-h_0(x)$ times the variance of $(Z \mid X > x)$, that is, as

$$-h_0(x)\{E(Z^2 \mid X > x) - [E(Z \mid X > x)]^2\}$$

which is negative.]

13. Suppose in a population, the force of mortality is age invariant, but there is variation between the forces of mortality of different individuals. If the force of mortality can be expressed in terms of a multiplicative frailty factor, show that a life table constructed without taking the frailty factor into account would show a force of mortality that decreases with age. Under the given setup, $h(x) = Z\lambda$, where λ is a constant. Hence $E[h(x) \mid X > x] = \lambda E(Z \mid X > x)$, which is a decreasing function of x (Exercise 12). The mechanism that brings this about may be that individuals with higher forces of mortality die at younger ages so that those with lower forces of mortality are selected to live longer.

14. Suppose in Population A, $h_A(t) = Z\lambda$, whereas in Population B, $h_B(t) = \mu\lambda$, where μ is the expected value of Z in Population A. Thus in Population A individuals vary in frailty, while in Population B there is no such variation. The force of mortality in the latter population is the former's mean force of mortality. Show that the life expectancy at birth in Population A is higher than that in Population B. In Population A the life expectancy at birth $= (1/\lambda)$ times the reciprocal of the harmonic mean of Z, whereas in Population B it is equal to $(1/\lambda)$ times the reciprocal of the arithmetic mean of Z. But the arithmetic mean is larger than the harmonic mean. Therefore the life expectancy in the heterogeneous population is higher than that in the homogeneous population.

15. Suppose there are N persons in a population with the force of mortality for the ith person being $h(x, i)$. Let $\lambda(x)$ be the mean (over individuals) of $h(x, i)$. Consider another population with homogeneous force of mortality $\lambda(x)$. Show that the survival function corresponding to $\lambda(x)$ is

$$\left[\prod_i S(x, i)\right]^{1/N}$$

where $S(x, i)$ is the survival function corresponding to $h(x, i)$. Also show that the survival function corresponding to $\lambda(x)$ is the median of $S(x, i)$'s when the distribution of $S(x, i)$ over individuals is symmetric (Nour, 1984).

Bibliographic Notes

1. *Modeling force of mortality for the whole life.* Reviews of attempts at finding a "law of mortality" (a mathematical formula to represent the age pattern of the force of mortality) are available in Elston (1923) and Benjamin and Haycocks (1970). The nineteenth century works of Gompertz (1825) and Makeham (1860) are considered to be the earliest attempts to find a simple parametric representation of the age pattern of the force of mortality. The Gompertz law has been found to work satisfactorily for the age groups 30 to 90 in most populations. Makeham's law may be thought of as a modification of the Gompertz's law via an additive term to represent accidental

deaths at younger ages. Attempts to represent parametrically the whole age pattern of the force of mortality have repeatedly used a component approach, involving, for example, three components, one to represent the risks of death in early life, one that in later life, and one that is "unnatural." Thus more than eleven decades ago Thiele (1872) proposed the model

$$\mu(x) = a_1 \exp(-b_1 x) + a_2 \exp[-\tfrac{1}{2} b_2 (x - c)^2] + a_3 \exp(b_3 x)$$

in which the last term is an increasing Gompertz curve to represent old age mortality, the first a decreasing Gompertz curve to represent the childhood mortality, and the middle term a normal curve that may be thought of as representing accident mortality superimposed on the "natural" mortality represented by the other two components. More recently Heligman and Pollard (1980) and Mode and Busby (1982) have proposed particular parametrization of the three components just mentioned. See Keyfitz (1981) for a review of the choice of function for mortality analysis. There is something to be said in favor of keeping the number of parameters to a minimum possible, if we are interested in using the models for forecasting.

2. *Mortality of the very old.*   As already mentioned, one of the old "laws" of mortality is the Gompertz curve, which has been found to fit observed mortality patterns between ages 30 and 90 rather well. The usual scantiness of observed data above age 90, and the generally small effect that mortality rates past age 90 have on life table calculations have prompted many actuaries to use without hesitation the Gompertz curve to extrapolate mortality to the end of life. Wilkin's (1981) study based on U.S. Medicare data has, however, questioned the wisdom of this practice.

3. *Heterogeneity in survival capacity of individuals.*   Vaupel et al. (1979) and Hougaard (1984a,b) among others have examined the effect of heterogeneity between individuals via a notion of frailty. (Discussions of heterogeneity can be found in Mode, 1985; Manton and Stallard, 1981a,b,1984; Keyfitz, 1984; Keyfitz and Littman, 1980; Vaupel and Yashin 1983, 1985; Manton et al., 1981a; Heckman and Singer, 1984; and Trussell and Richards, 1985.)

Chapter 4 | Life Tables Based on Survey or Observational Data

This chapter is concerned with life tables based on survey or observational data. Examples of the former include life tables of birth intervals, contraceptive practices, marriage history, and migration history, the basic data being drawn from information collected in sample surveys or censuses. Examples of life tables based on "observational" data are life tables showing the failure times of small business establishments and those showing the disbandment or dissolution times of radical political organizations. Survey data are usually collected prospectively or retrospectively. In a prospective (longitudinal or panel) study, an aggregate of individuals or units is followed up in order to obtain the needed information over time. In the retrospective approach, the subjects selected for study are asked to report their past experiences. Patient files maintained on a continuous basis by clinics and population registers that record changes in status of individuals as they occur provide prospective data. Censuses and sample surveys often collect retrospective information by asking questions such as "When did your first marriage begin?," "When was your first child born?," etc.

1 Nature of Data

Irrespective of whether the data are derived from prospective or retrospective studies, ideally, the life table construction requires information on the following dates:

1. Date of entry (date of randomization)
2. Failure date (date of experiencing the event of interest)

3. Date of exit from the study
   a. Date of censoring
4. Stopping date (cutoff date) [last follow-up date in clinical studies; interview date in surveys; end of observation period in observational studies]

The date of entry theoretically marks the beginning of exposure of the individual or unit under study to the risk of experiencing the event of interest. Examples are birth date in mortality studies, the date of marriage in marriage dissolution studies, and the date of nth move in migration studies. Illustrative of the date of experiencing the event of interest are the date of death in mortality studies, the date of $(n + 1)$th move in migration studies, and the date of separation or divorce in marriage dissolution studies. The date of exit from a study stands for the date when the unit under observation ceases to be under observation, for whatever reason. The stopping date stands for the last follow-up date in surveys and for the end of the chosen observation period in observational studies involving secondary data (e.g., a fixed date such as December 31, 1984 in a study of *coups d'etat* in independent nations).

Note that the date of entry corresponds to the date of randomization (that is, the time when subjects are randomly assigned to different treatment regimens) in experiments. Also note that the date of experiencing the event of interest is more commonly referred to, in the context of experiments, as the date of failure, an alternative term being the date of termination, i.e., the date that marks the end of exposure to the risk of experiencing the relevant event. We shall use the terms date of failure and date of termination interchangeably. Correspondingly, failure time and duration to termination will be regarded as interchangeable terms.

If an individual does not experience the relevant event before the date of exit (or the stopping date as the case may be), his (her) record is treated as censored. In such cases all that is known is that failure had not occurred till the date of exit (stopping date).

The inclusive term "trial time" is sometimes used for failure and censoring times together (Peto *et al.*, 1977). Recorded exposure time may be an alternative label for trial time.

2   Mechanics

Suppose we are given the dates of entry, termination, and interview for a sample of individuals (see Table 4.1). From these dates we calculate the durations to termination and to interview by counting the units of time elapsed between the pertinent dates. In Table 4.1, these durations have been

Table 4.1

Hypothetical Survey Data for the Construction of a Life Table[a]

Case	Date of entry	Date of termination[b]	Date of interview	Duration to termination[b]	Duration to interview
a	Jan 2	Feb 11	May 25	40	143
b	Jan 17	May 4	May 17	107	120
c	Jan 18	—	May 10	—	112
d	Jan 22	Feb 28	May 13	37	111
e	Feb 10	May 17	May 23	96	102
f	Jan 30	Feb 12	May 15	13	105
g	Apr 4	—	May 6	—	32
h	Apr 29	—	May 27	—	28
i	May 18	—	May 29	—	11
j	May 20	—	May 31	—	11
k	May 15	—	May 18	—	3
l	Feb 5	Feb 25	May 19	20	103
m	Feb 5	Apr 18	May 10	72	94
n	Feb 6	May 18	May 28	101	111
o	Feb 26	—	May 22	—	85
p	Mar 10	—	May 25	—	76
q	Mar 11	May 8	May 12	58	62
r	Mar 28	—	May 29	—	62
s	Mar 15	Mar 23	May 10	8	56
t	Apr 13	—	May 20	—	37
u	Apr 4	May 9	May 11	35	37
v	Apr 25	May 16	May 31	21	36

[a] All events are assumed to have taken place in one calendar year (e.g., 1986).
[b] — indicates censoring

calculated in days. [See below for a few comments on the use of other units of time.]

In Table 4.2, the cases presented in Table 4.1 are arranged in ascending order of trial time, with cases experiencing the relevant event during day T, i.e., those with failure time T days, appearing just before those with censoring time T days (see, e.g., cases d and t with $T = 37$).

Column 2 of Table 4.2 shows the serial numbers for the cases starting at the bottom of the table and going upwards. The highest serial number, which goes to the case appearing in the first row, is equal to the sample size (22, in the present instance). This column is not essential. It is introduced here as a helpful device in filling out a later column.

Column 3 of Table 4.2 indicates for each case whether the relevant event was experienced (y for "yes" and "n" for no). The trial times are shown in Column 4, using brackets to indicate ties (see, e.g., $T = 11$ and $T = 37$). All subsequent columns of the table pertain to distinct trial times.

Table 4.2

Estimation of Survival Function

Case	Serial number	Was the event experienced? yes = y no = n	Trial time t_i	No. of events at t_i	No. of cases with trial time $> t_i$	P = "Obs." survival proportions[a]	Estimate of survival function[b]
(1)	(2)	(3)	(4)	(5)	(6)	(7)	(8)
k	22	n	3				
s	21	y	8	1	21	0.95238	0.95238
i	20	n	11⎤				
j	19	n	11⎦				
f	18	y	13	1	18	0.94444	0.89947
l	17	y	20	1	17	0.94118	0.84656
v	16	y	21	1	16	0.93750	0.79365
h	15	n	28				
g	14	n	32				
u	13	y	35	1	13	0.92308	0.73260
d	12	y	37⎤	1	12	0.91667	0.67155
t	11	n	37⎦				
a	10	y	40	1	10	0.90000	0.60440
q	9	y	58	1	9	0.88889	0.53724
r	8	n	62				
m	7	y	72	1	7	0.85714	0.46049
p	6	n	76				
o	5	n	85				
e	4	y	96	1	4	0.75	0.34537
n	3	y	101	1	3	0.66667	0.23025
b	2	y	107	1	2	0.5	0.11512
c	1	n	112				

[a] Column 7 = 1 − [(Col. 5)/(Col. 6)]. Thus $0.95238 = \frac{20}{21}$.

[b] Column 8 = the continued product of the figures in Column 7 until the trial time in question. Thus $0.89947 = 0.95238 \times 0.94444$.

Column 5 shows the number of events occurring during each day, and column 6 shows the number of cases at risk during each day. The serial numbers in column 2 help to fill this column—one simply transfers the relevant serial numbers from column 2 to column 6, remembering that when there are ties only the serial number of the top entry is transferred. Column 6 can also be filled out noting that at the beginning the number at risk is equal to the sample size and that this number is decreased over time in accordance with the number of cases that experience the relevant event or are censored. Thus, since one case is censored on day 3, the number at risk at the end of that day is equal to the corresponding number as of the end of the previous day less one, i.e., $22 - 1 = 21$. No further change in the number at risk occurs until day 8. Since one case experiences the events of interest on that day, the number at risk as of the end of that day (or, which is the same thing, as of the beginning of the next day) is 20, and so on.

Column 7 of the table is obtained by dividing column 5 by column 6 and then subtracting the result from unity. The entry in this column for day T gives the observed proportion P of individuals completing that day without experiencing the relevant event, among, of course, those who have not had the experience as of the beginning of the day.

Column 8 of the table is obtained for day T by calculating the continued product of the entries in column 7 up to and including the one for day T. These numbers regarded as a function of T represent the estimated survival function of the life table. Thus according to the numbers shown in the column, the estimated probability of remaining free from experiencing the event for the first 58 days of exposure is 0.53724, that of completing 107 days without experiencing the event is 0.11512, and so on.

If one graphs the survival function estimated in the manner just described, spurious jumps or long flat regions may sometimes occur in the plot. For example, the graph of the survival function in Table 4.2 is flat from day 8 to day 13, from day 21 to day 35, from day 72 to day 96, and so on. This is because the data are so sparse that no one happens to have experienced the relevant event during these periods. Conclusions based on the fine details of such a plot are likely to be wrong. For example, the flat region from day 72 to day 96 in Table 4.2 does not necessarily mean that individuals who have not had the experience of the event until day 72 will remain in that state for another 24 days. To warrant such inferences one should have a very large number of individuals whose failure times exceed the values of T corresponding to the flat portion in question.

If the data show that during each unit interval of time several individuals experience the event of interest or are censored, then the calculations proceed essentially in the same manner as shown in Table 4.2; each individual case is not, however, accorded a separate line.

The following points are worth emphasizing about the procedure described above.

Remark 1. Obviously, in constructing Table 4.2 the experiences of a number of persons with differing entry dates are put together so as to yield a picture concerning the experiences of a synthetic cohort with a common entry date.

Remark 2. There are two fundamental assumptions involved: The population is homogeneous, and the censoring and failure mechanisms are independent. A population is said to be homogeneous if and only if all members of the population are subjected to the same conditional probability of experiencing the event under study at duration T, given that they are at risk (of experiencing the event) as of time T. The independence of the failure and censoring mechanisms implies, for example, that individuals are not selectively censored because of a relatively poor or relatively good prognosis.

Remark 3. In the illustration given in Table 4.2, trial times are calculated in days. The choice of the time is dictated by the availability of details in the data, their accuracy, and the rapidity with which the relevant event occurs after the time of entry. If, for example, the analyst knows only the month and year of the various dates, it is meaningless to calculate trial times in days. If there is reason to believe that the recorded day of the month of termination or of entry is inaccurate in a large number of cases, there is little point in measuring trial times in days. On the other hand, if the relevant event tends to occur with high frequency within a few days after the date of entry, then the use of a month or year for measuring time may give only a crude picture of the early part of the survival function.

Remark 4. When we say that the trial time is T, we mean that it is greater than or equal to T but less than $T + 1$. A notation that makes this explicit is $[T, T + 1)$. Often we distinguish between exact trial time and grouped trial time. Suppose, for example, the dates (day, month, year) of marriage and first birth are available permitting the calculation of the interval between marriage and first birth in days. Usually such information is referred to as exact time. The label is appropriate, given that the unit of time used is a day. If, on the other hand, one were to regard hour as the unit of time, then the information just mentioned in terms of days is to be viewed as grouped information, since we know only that the event (first birth) of interest occurred sometime during an interval of several units in length.

Remark 5. The calculation of the P values in column 7 of Table 4.2 is based on the assumption that censoring occurs at the end of the time interval involved. Thus, case t is assumed to have been censored at the end of day 37. The impact of the violation of this assumption on the estimate of the survival function can be appreciated by noting that if case t was censored at the beginning of day 37 (which is the same as the end of day 36), the number at risk during day 37 would have been 11 instead of 12. Practice varies with respect to the way of handling the censored cases; some practitioners use small units to measure trial times and assume that censoring occurs at the end of the intervals. Some experts advocate the assumption that censoring occurs at the beginning of an interval. Some favor the assumption that censoring times are evenly distributed over the time interval involved. If the unit used for measuring time is small, these procedures produce practically the same result. When the units are large, however, different assumptions may lead to different estimates of the survival function. Let us examine this matter in some detail.

3   Different Ways of Handling Censoring Times

Suppose, referring to Table 4.1, durations to interview and terminations are given in months instead of days. Taking 30 days as one month, we have then the available information in the form shown in Table 4.3.

Note that for cases u, e, and n, the termination month coincides with the interview month. Note also that for i, j, k, and h, the interview month is the same as the month of entry. If we assume that the interviews took place at the beginning of the month indicated, then none of this quartet (i, j, k, h) should be regarded as at risk during $[0, 1)$, implying that 18 is the number at risk during $[0, 1)$. Similarly, the number at risk during $[1, 2)$ is 11; and so on. A general way of calculating the number at risk during $[T, T + 1)$ is to add the following quantities:

a. the number of individuals whose trial times are greater than or equal to $T + 1$, and

b. the number of terminations in $[T, T + 1)$, excluding individuals with the time to interview also falling in $[T, T + 1)$.

Table 4.3

Data in Table 4.1 Recast Using a Month ($= 30$ days) as the Time Unit

Trial time	Cases terminating at t_i and interviewed later	Cases terminating at t_i and interviewed at t_i	Cases interviewed at t_i, but with no termination until then
0	s, f, l, v		k, i, j, h
1	d, a, q	u	g, t
2	m		r, p, o
3	b	e, n	c

The exclusion of terminations that occur in the same interval as the interview is consistent with the assumption that censoring occurs at the beginning of the interval and the rule that terminations that occur after censoring are to be ignored in the analysis.

Table 4.4 gives the estimated survival function corresponding to the information in Table 4.3, assuming that censoring occurred at the beginning of the pertinent intervals.

The estimates shown in Table 4.4 may be compared with those given in Table 4.5, the latter based on the assumption that censoring occurred at the end of the interval indicated. [Note that whereas the terminations for cases u, e, and n are ignored in the calculations in Table 4.4, they are counted in Table 4.5.]

Put differently, if n is the number of individuals surviving to the beginning of interval $[T, T + 1)$, including those who are censored in the interval, then in Table 4.5 the formula used for the calculation of P is

$$P = 1 - \text{terminations in } [T, T + 1)/n \qquad (4.1)$$

Table 4.4

Survival Function Estimated from Table 4.3[a]

Trial time	Number of terminations	Number at risk	P	Estimated survival function
0	4	18	14/18	14/18
1	3	11	8/11	(14/18)(8/11)
2	1	5	4/5	(14/18)(8/11)(4/5)
3	1	1	0	0

[a] Cases k, i, j, and h (see Table 4.3) are ignored from the calculations, the month of interview for them being the same as the month of entry. It is assumed that interview took place at the beginning of the month.

Table 4.5

Another Analysis of the Data in Table 4.3

Trial time	Terminations	Number at risk	P	Estimated survival function
0	4	22	18/22	18/22
1	4	14	10/14	(18/22)(10/14)
2	1	8	7/8	(18/22)(10/14)(7/8)
3	3	4	1/4	(18/22)(10/14)(7/8)(1/4)

whereas the corresponding formula used in Table 4.4 is

$$P = 1 - \frac{\text{terminations in } [T, T + 1)}{n - \text{censorings in } [T, T + 1)} \tag{4.2}$$

A compromise between these is the following

$$P = 1 - \frac{\text{terminations in } [T, T + 1]}{n - a\{\text{censorings in } [T, T + 1)\}} \tag{4.3}$$

where a is the average fraction of the interval for the censored cases.

4 The Reduced Sample Method

All of the procedures discussed so far involve estimating the conditional probability of survival for each interval and then using their continued products to estimate the survival function. A procedure called the reduced sample method can be used to estimate the survival function more directly. The formula for this is the following:

$$\hat{S}(T + 1) = \frac{n_{T+1}}{n_{T+1} + \text{terminations in } [T, T + 1)} \tag{4.4}$$

where $\hat{S}(T + 1)$ is the estimated probability of survival until $T + 1$, and n_{T+1} the number of individuals surviving until $T + 1$. Table 4.6 illustrates the application of this method to the data in Table 4.3.

5 Product–Limit Estimate of the Survival Function

Suppose all of the individuals are kept under observation until the termination event occurs. If $t_1, \ldots, t_n$ are the exact survival times of the n individuals in the sample, an estimate of the survival function $S(t)$ is the

Table 4.6

Application of the Reduced Sample Method to the Data in Table 4.3

Trial time	Number of terminations at t_i	Number of cases with trial time $> t_i$	Estimated survival function
0	4	22	$14/(14 + 4)$
1	4	14	$8/(8 + 4)$
2	1	8	$4/(4 + 1)$
3	3	4	0

observed proportion of individuals in the sample who survive longer than t; i.e.,

$$\text{estimate of } S(t) = \frac{\text{number of individuals who survive longer than } t}{\text{total number of individuals in the sample}} \qquad (4.5)$$

If the observed survival times $t_1, \ldots, t_n$ are ordered in ascending order and if they are labeled, say,

$$t_{(1)} < t_{(2)} < \cdots < t_{(n)}$$

then the survival function at $t_{(i)}$ can be estimated as

$$\text{estimate of } S(t_{(i)}) = (n - i)/n \qquad (4.6)$$

where $(n - i)$ is the number of individuals in the sample surviving longer than $t_{(i)}$. If ties occur, the largest "i" value is used in applying this formula. Thus if $t_{(3)} = t_{(4)} = t_{(5)}$ then $(n - 5)/n$ is our estimate of the values of the survival function at $t_{(3)}$, $t_{(4)}$, and $t_{(5)}$.

Clearly the estimated survival function is a step function starting from 1 at $t = 0$ and reaching 0 at $t = \infty$. If there are no ties, then the decrease at each stage is $1/n$ in magnitude. The accompanying table is a simple illustration of the calculations involved.

Trial time $t_{(i)}$	3	4	5	5	5	6	10	10	12
Number of cases with trial time $> t_{(i)}$	8	7	4	4	4	3	1	1	0
Estimated survival function	$\frac{8}{9}$	$\frac{7}{9}$	$\frac{4}{9}$	$\frac{4}{9}$	$\frac{4}{9}$	$\frac{3}{9}$	$\frac{1}{9}$	$\frac{1}{9}$	0

If some of the subjects do not experience the event under study any time before the study ends, a method different from the one illustrated above is obviously required. A very commonly used procedure is known as the Kaplan–Meier (1958) product-limit (PL) method. Table 4.7 illustrates the calculations involved. Notice the similarities in the procedure illustrated in Tables 4.7 and 4.2.

A few remarks about the PL estimates of the survival function are in order. [We shall use for convenience the notation $s(t)$ for an estimate of $S(t)$.]

Remark 1. It is easy to see that

$$s(t_{(i)}) = s(t_{(i-1)})(n - i)/(n - i + 1) \qquad (4.7)$$

where $t_{(i)}$ and $t_{(i-1)}$ are uncensored observations. Thus, from Table 4.7, $s(9.0) = s(7.0)(\frac{1}{2})$.

Remark 2. If there are no censored observations

$$s(t_{(r)}) = \frac{n-1}{n} \frac{n-2}{n-1} \cdots \frac{n-r}{n-r+1} \qquad (4.8)$$

Table 4.7

The Product Limit Method of Estimating the Survival Function

Trial time[a]	Rank[b]	Rank of uncensored observations[c]	Est. of the conditional prob. of survival[d]	Product-limit estimate of the survival function[e]
2.3	1	1	$\frac{9}{10}$	$\frac{9}{10}$
4.1+	2			
5.5+	3			
5.9	4	4	$\frac{6}{7}$	$(\frac{9}{10}) \times (\frac{6}{7})$
5.9	5	5	$\frac{5}{6}$	$(\frac{9}{10}) \times (\frac{6}{7}) \times (\frac{5}{6})$
6.3+	6			
7.0	7	7	$\frac{3}{4}$	$(\frac{9}{10}) \times (\frac{6}{7}) \times (\frac{5}{6}) \times (\frac{3}{4})$
7.0+	8			
9.0	9	9	$\frac{1}{2}$	$(\frac{9}{10}) \times (\frac{6}{7}) \times (\frac{5}{6}) \times (\frac{3}{4}) \times (\frac{1}{2})$
11.0	10	10	0	0

[a] Censored observations are indicated by a "+". If T is a termination time and $T+$ is a censored time, then T is entered first followed by $T+$.

[b] The second column shows the rank (the line number) of the observations in column 1.

[c] The third column repeats the entries in the second column, for the uncensored cases.

[d] The fourth column shows the values of $(n - r)/(n - r + 1)$, where $n = 10$, and r is the entry in column 3.

[e] The last column gives the continued product of the entries in the last-but-one column.

which simplifies to $(n - r)/n$. Note that this is the same as the formula given earlier in this section for the situation in which all cases are followed up until everyone experiences the event under study.

Remark 3. The mean survival time can be estimated as the mean of the PL estimates. It is equal to the area under the survival curve. If the longest time observed is a censored observation, then the mean survival time is not usually computed.

Remark 4. An estimate of the variance of the PL estimate is

$$[s(t_{(r)})]^2 \sum_{i=1}^{r-1} \frac{1}{(n - i)(n - i + 1)} \tag{4.9}$$

Thus from Table 4.7, an estimate of the variance of $s(9.0)$ is

$$[s(9.0)]^2 \left[\frac{1}{9 \times 10} + \frac{1}{6 \times 7} + \frac{1}{5 \times 6} + \frac{1}{3 \times 4} \right]$$

The square root of the estimated variance can be used for the computation of confidence intervals for the estimated survival function.

6   An Example of Life Table Construction with Grouped Data

We now present an example using data pertaining to remarriage for women of age 25 to 34. The original data are from the 1975 U.S. Current Population Survey. The basic information from the data source and some of the preliminary calculations are shown in Table 4.8.

Column 1 shows the time interval used for grouping the observations, all intervals being of length one year, except the last one, which is kept open. Column 2 shows the number remaining divorced as of the beginning of the time interval. Column 3 shows the number of remarriages occurring during the time interval. Column 4 shows the number of women remaining divorced as of the interview date. Column 5 shows the number of individuals at risk of remarriage at the beginning of the interval. Column 6 gives estimated probability of remarriage during the interval, conditional on remaining divorced at the beginning of the interval. Column 7 gives $p_i = 1 - q_i$, $i = 1$, 2, ... Thus, $p_1 = 0.8092$, $p_2 = 0.8090$, and so on.

Table 4.8

Data and Preliminary Calculations: A Life Table for Remarriage; United States, 1975

Interval	n_i'	d_i	c_i	n_i	q_i	p_i
(1)	(2)	(3)	(4)	(5)	(6)	(7)
[0, 1)	1298	238	101	1247.5	0.1908	0.8092
[1, 2)	959	177	65	926.5	0.1910	0.8090
[2, 3)	717	107	51	691.5	0.1547	0.8453
[3, 4)	559	64	42	538.5	0.1190	0.8810
[4, 5)	453	50	34	436.0	0.1147	0.8853
[5, 6)	369	34	18	360.0	0.0944	0.9056
[6, 7)	317	27	21	306.5	0.0881	0.9119
[7, 8)	269	21	11	263.5	0.0797	0.9203
[8, 9)	237	14	4	235.0	0.0596	0.9404
[9, 10)	219	15	11	213.5	0.0703	0.9297
[10, 11)	193	16	14	186.0	0.0860	0.9140
[11, 12)	163	8	8	159.0	0.0503	0.9497
[12, 13)	147	6	8	143.0	0.0420	0.9580
[13, 14)	133	6	7	129.5	0.0463	0.9537
[14, 15)	120	5	5	117.5	0.0426	0.9574
[15, 16)	110	2	10	105.0	0.0190	0.9810
[16, 17)	98	5	5	95.5	0.0524	0.9476
[17, 18)	88	5	2	87.0	0.0575	0.9425
[18, 19)	81	2	2	80.0	0.0250	0.9750
[19, 20)	77	2	9	72.5	0.0276	0.9724
[20, ∞)	66	12	54	39.0	0.3077	0.6923

Table 4.9 shows estimated survival, hazard, and probability density functions, and the corresponding estimated standard errors of estimates. The survival function is estimated by computing the continued products of the p_i's. Thus, $s(1) = p_1$, $s(2) = p_1 p_2$, and so on. For convenience of notation, we define $p_0 = 1$, so that we have for the ith interval ($i = 1, 2, \ldots$)

$$s(i) = p_0 p_1 \cdots p_{i-1} p_i, \qquad i = 0, 1, \ldots \tag{4.10}$$

An estimate of the variance of $s(i)$ is the square of $s(i)$ multiplied by one less than the continued product of $[1 + (1/n_i)(q_i/p_i)]$. Thus, $\mathrm{var}[s(0)] = 0$, $\mathrm{var}[s(1)] = (1/n_1)(q_1/p_1)[s(1)]^2$, and so on. A convenient computational formula is given by Greenwood's approximation $[s(t)]^2 \Sigma (1/n_i)(q_i/p_i)$, where the summation is up to t. From the estimated survival function, we estimate the values of t for which $s(t) = 0.75, 0.5,$ and 0.25 as 1.38, 3.81, and 11.92,

Table 4.9

Estimates of Survival, Hazard, and Probability Density Functions and Standard Errors of the Estimates for the Life Table Constructed, Based on the Data in Table 8

t	$s(t)$	SE[$s(t)$]	$h(t)^a$	Standard error [$h(t)$]	pdf^a	Standard error (pdf)
(1)	(2)	(3)	(4)	(5)	(6)	(7)
0	1.0000	0.0	0.2109	0.0136	0.1908	0.0001
1	0.8092	0.0111	0.2112	0.0158	0.1546	0.0001
2	0.6546	0.0138	0.1677	0.0162	0.1013	0.0001
3	0.5533	0.0147	0.1265	0.0158	0.0658	0.0001
4	0.4875	0.0151	0.1217	0.0172	0.0559	0.0001
5	0.4316	0.0153	0.0991	0.0170	0.0408	<0.00005
6	0.3908	0.0154	0.0922	0.0177	0.0344	<0.00005
7	0.3564	0.0154	0.0830	0.0181	0.0284	<0.00005
8	0.3280	0.0154	0.0614	0.0164	0.0195	<0.00005
9	0.3085	0.0153	0.0728	0.0188	0.0217	<0.00005
10	0.2868	0.0152	0.0899	0.0224	0.0247	<0.00005
11	0.2621	0.0151	0.0516	0.0182	0.0132	<0.00005
12	0.2489	0.0150	0.0429	0.0175	0.0104	<0.00005
13	0.2385	0.0150	0.0474	0.0194	0.0110	<0.00005
14	0.2274	0.0150	0.0435	0.0194	0.0097	<0.00005
15	0.2178	0.0149	0.0192	0.0136	0.0041	<0.00005
16	0.2136	0.0149	0.0538	0.0240	0.0112	<0.00005
17	0.2034	0.0150	0.0592	0.0265	0.0116	<0.00005
18	0.1908	0.0150	0.0253	0.0179	0.0048	<0.00005
19	0.1860	0.0150	0.0280	0.0198	0.0051	<0.00005
20	0.1809	0.0150	—	—	—	—

^{a} The hazard function and the probability density function were not estimated for the open interval 20+.

respectively. See Figs. 4.1, 4.2, and 4.3 for graphical representations of the estimated survival, hazard, and density functions.

It may be noted that from Table 4.9 that we could estimate the probability of remarriage within a given number of years after divorce. Thus we estimate the probability of remarriage within five years of divorce as $1 - s(5) = 1 - 0.4316 = 0.5684$. Similarly, $s(20) = 0.1809$ indicates that almost 82 percent of women remarry within 20 years of divorce. Among women remarrying within 20 years of divorce, the median time to remarriage is the value of t for which $[1 - s(t)]$ equals one-half of $[1 - s(20)]$, or $s(t)$ equals one-half of $[1 + s(20)]$.

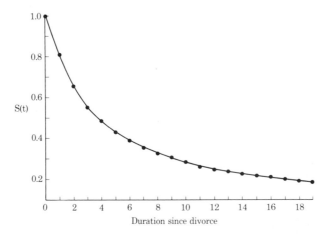

Fig. 4.1. Estimated survival function: Remarriage, women age 25–34, U.S., 1975.

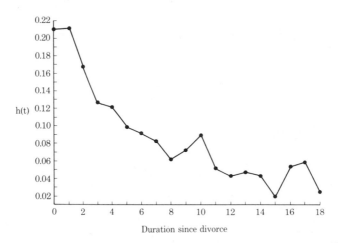

Fig. 4.2. Estimated hazard function: Remarriage, women age 25–34, U.S. 1975.

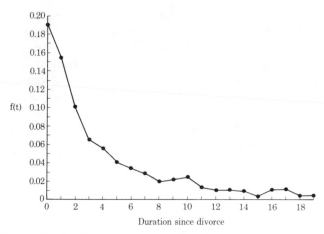

f(t)

Duration since divorce

Fig. 4.3. Estimated probability density function, remarriage, women age 25–34. U. S., 1975.

Problems and Complements

1. Freireich *et al.* (1963) report a clinical trial in which a drug 6-MP (6-mercaptopurine) was compared with a placebo with respect to maintaining remission in leukemia patients. The following are the remission times for the experimental and the control groups. The censoring times are marked with a "+".

Experimental group: 6, 6, 6, 6+, 7, 9+, 10, 10+, 11+, 13, 16, 17+, 19+, 20+, 22, 23, 25+, 32+, 32+, 34+, 35+
Control group: 1, 1, 2, 2, 3, 4, 4, 5, 5, 8, 8, 8, 8, 11, 11, 12, 12, 15, 17, 22, 23

---

   a. Compute the PL estimate of the survival functions for the two groups.
   b. Estimate the variance of $s(t)$ for each uncensored observation.
   c. Estimate the median survival time of the two groups.
   d. Estimate the first and third quartiles of the survival times of the two groups.

2. The data in Table 4.10 are from Rodriguez and Hobcraft (1980).

By duration of exposure is meant the interval between marriage to the first birth or interview, whichever occurs first. The term termination status is used for a three category variable indicating whether exposure was terminated by (i) the interview, (ii) the interview and the first birth, or (iii) the first birth.

   a. Treat the categories (i) and (ii) of the termination status as censoring and complete the life table.
   b. Compute the first, second, and third quartiles of the survival function.
   c. Estimate the proportions of women having a birth by duration x among women who have a birth within five years of marriage. Calculate the first, second, and third quartiles of this "standard" distribution.

Table 4.10

Duration of exposure (months)	Number observed at start	Termination status		
		Interview	First birth and interview	First birth
[0, 3)	3302	16	5	394
[3, 6)		28	0	119
[6, 9)		52	3	227
[9, 12)		26	4	793
[12, 15)		16	1	389
[15, 18)		7	4	229
[18, 21)		8	2	175
[21, 24)		5	1	185
[24, 27)		4	1	104
[27, 30)		4	0	68
[30, 33)		3	0	51
[33, 36)		5	0	54
[36, 39)		4	0	35
[39, 42)		2	0	29
[42, 45)		5	0	16
[45, 48)		1	0	21
[48, 51)		5	0	21
[51, 54)		0	0	15
[54, 57)		4	0	8
[57, 60)		2	0	9
[60, 63)		2	0	8
[63, 66)		1	0	7
[66, 69)		1	0	5
[69, 72)		6	0	6
[72, 75)		2	0	5
[75, 80)		48	51	0

d. Compute Tukey's (1978) trimean given by Trimean = $(Q_1 + 2Q_2 + Q_3)/4$ where Q_1, Q_2, and Q_3 are the first, second, and third quartiles.

3. The following figures are from Cutler *et al.* (1960a).

Time from diagnosis (years)	Number lost to follow-up	Number withdrawn alive	Number dying
[0, 5)	18	0	731
[5, 10)	16	0	52
[10, 5)	8	67	14
[15, 20)	0	33	10

a. Given that the sample was of size 949, compute the life table showing

    i. n_i', the number of patients entering the ith interval alive.

    ii. n_i, the number of individuals exposed to risk in the ith interval (use the formula $n_i = n_i' - \frac{1}{2}$ (number lost to follow-up and number withdrawn).

    iii. q_i, the conditional probability of dying in the ith interval, given exposure to the risk of dying in that interval.

    iv. $p_i = 1 - q_i$.

    v. an estimate of the survival function.

    vi. an estimate of of the probability density function.

    vii. An estimate of the hazard function. [HINT: Let $s(t_i)$ be your estimate of the survival function at time t_i. This is your estimate of the proportion surviving until t_i. The ordinate of the probability density function at t_{mi}, midway between t_i and t_{i+1}, is estimated by dividing $s(t_i) - s(t_{i+1})$ by $t_{i+1} - t_i$. If $s(t_{mi})$ denotes the average of $s(t_i)$ and $s(t_{i+1})$, then the ordinate of the hazard function at t_{mi} is estimated as the ratio of the corresponding ordinate of the probability density function to $s(t_{mi})$.]

4. The large-sample approximate variance of $h(t_{mi})$, the estimated ordinate of the hazard function at t_{mi} (Gehan, 1969), is

$$\frac{[h(t_{mi})]^2}{n_i q_i}\left[1 - \frac{[w_i h(t_{mi})]^2}{4}\right]$$

where $w_i = t_{i+1} - t_i$. Apply this formula to obtain the variance of the estimated hazard function in Exercise 3.

5. Referring to the life table constructed in Exercise 2, let $[t_j, t_{j+1}]$ be the interval such that $s(t_j) \leq 0.5$ and $s(t_{j+1}) < 0.5$. Then the median survival time can be estimated as

$$t_j + \frac{[s(t_j) - 0.5]w_j}{s(t_j) - s(t_{j+1})}$$

where $w_j = t_{j+1} - t_j$. What formula would you use to obtain the first and third quartiles of the survival function?

6. Note that the median remaining lifetime at t_i is estimated as the value on the time axis corresponding to which the value of the survival function is $\frac{1}{2}s(t_i)$. If $[t_j, t_{j+1})$ is the interval such that $s(t_j) \geq 0.5s(t_i)$ and $s(t_{j+1}) < 0.5s(t_i)$, then the median remaining lifetime at t_i is

$$(t_j - t_i) + \frac{(t_{j+1} - t_j)(s(t_j) - 0.5s(t_i))}{s(t_j) - s(t_{j+1})}$$

Using this formula, estimate the median remaining lifetime at $t = 21$ months in the life table you constructed in Exercise 2.

7. The product-limit estimate can be defined as follows: Suppose there are observations on n individuals and that there are $k(\leq n)$ distinct times, say, $t_{(1)} < t_{(2)} < \cdots < t_{(k)}$ at which failures occur. Let $d_{(i)}$ represent the number of failures at $t_{(i)}$. Then the product limit estimate of $S(t)$ is defined as

$$s(t) = \prod \frac{n_{(j)} - d_{(j)}}{n_{(j)}}$$

where $d_{(j)}$ is the number of deaths at t_j, $n_{(j)}$ is the number of individuals at risk at $t_{(j)}$ and the product is taken over j such that $t_{(j)} < t$. [The number of individuals at risk at t is the number of individuals alive and uncensored as of the moment just prior to t.] When applying this formula, individuals with censoring times equal to t_j are included in n_j.

Compute the product-limit estimate of the survival function given the accompanying data.

$t_{(j)}$	$n_{(j)}$	$d_{(j)}$	$s(t_j)$
6	21	3	0.857
7	17	1	0.807
10	15	1	
13	12	1	
16	11	1	
22	7	1	
23	6	1	

8. From the following trial times (+ indicates censoring) prepare a table showing $t_{(j)}, n_{(j)}, d_{(j)}$, and $s(t_j)$.

5, 5, 5, 5+, 8, 9+, 10, 10+, 12+, 13, 15, 16+, 18+, 20+, 22, 24, 26+, 30+, 32+, 35+, 36+

Note that if the largest observed trial time is a censoring time, then the PL estimate is taken as defined up to the largest failure time. Thus, in the numerical example given above the PL estimate is taken as being defined only up to 24.

9. Plot the estimated survival function of Exercise 8, and mark a confidence band for the estimated survival function.

Bibliographic Notes

The product-limit approach to the estimation of the survival function was proposed by Böhmer (1912). Actuarial estimates, of course, were in use much earlier. Kaplan and Meier (1958) supplied a maximum likelihood basis for the product limit estimator. See also Johansen (1978), Efron (1967), and Cox and Oakes (1984). From a large-sample perspective, the life table and product-limit estimates have been examined under random censorship by Breslow and Crowley (1974). For a discussion using martingale theory see Aalen (1976). Other pertinent works include Altschuler (1970), Efron (1981), Meier (1975), Reid (1981a,b), Thomas and Grunkemeier (1975) and Turnbull (1976).

The beginning of the application of life table methods to the analysis of demographic surveys and observational studies can be traced to a number of publications that appeared in the 1950s and 1960s, among which are Brass (1958), Tuan (1958), Dandekar (1963), Potter (1963, 1967, 1969), Sheps (1965), and Wolfers (1968). Potter and Sheps are believed to have pioneered the application of life table techniques in the analysis of fertility and family planning data. A number of expository papers on the use of life table methods for the analysis of data from fertility surveys have been recently published under the auspices of the World Fertility Survey (see, e.g., Smith, 1980; Rodriguez and Hobcraft 1980; Hobcraft and Rodriguez, 1980; Rodriguez, 1984). Trussell and Hammerslough (1983) have fitted to the Sri Lankan Fertility Survey data a hazards model revealing the covariates of infant and child mortality. In that publication, the authors present a detailed discussion of hazard models in a language familiar to demographers. Rodriguez *et al.* (1983) present a comparative analysis of birth intervals using World Fertility Survey data for a number of countries. Trussell and Menken (1982) demonstrate how to use life table methods to analyze contraceptive failure. Martin *et al.* (1983) have examined the covariates of child mortality by fitting hazard models to the World Fertility Survey data from the Philippines. Indonesia, and Pakistan.

A number of special problems that arise in the construction of life tables from clinical records have been examined by Tietze and Lewit (1973) and Jain and Sivin (1977).

Chapter 5 | Statistical Comparison of Life Tables

Demographers are well aware that different population groups have different age patterns of mortality. Thus from an empirical analysis of a large number of life tables, Coale and Demeny (1966) identified four distinct patterns, one pertaining to Northern European countries, one to Southern European countries, one to Eastern European countries, and one to Western European populations and their overseas offshoots. Preston (1976), on the basis of an analysis of the association between age patterns of mortality and the cause-of-death structure, came to the conclusion that, in addition to the four patterns identified by Coale and Demeny, there exists at least one definite non-Western pattern, almost exclusively pertaining to Latin American countries. Others who have examined data from developing countries have concluded that a number of age patterns of mortality, which systematically differ from those identified by Coale and Demeny and Preston, do prevail in modern populations. The question naturally arises as to how to test, given two or more age patterns of mortality, whether they are different from one another. Going beyond the death process and letting the concept of survival time represent the duration in which a unit, such as an individual, a nation, or business corporation, remains free from a specific event such as marriage, mobility, promotion, going out of business, or a coup d'etat, the question posed above can be put in a more general form as: How does one test whether the survival functions underlying two or more samples of survival times are similar?

Often no formal statistical tests are needed to compare two or more samples of survival times; a visual inspection of the life table graphs will reveal the story. But frequently it may be difficult or impossible to tell whether the difference between two life table graphs are significant or whether it simply reflects chance fluctuations. Hence there is a need to be familiar with statistical tests that can be applied for comparing patterns underlying two or more samples of survival times.

This chapter outlines a number of nonparametric tests that have been developed for use in life testing and clinical trials. The section that follows is devoted to a brief discussion of certain data features. After that two-sample comparisons and k (≥ 2) sample problems are discussed. The chapter closes with a demonstration of the use of poststratification in life table comparisons. As in previous chapters a number of topics related to those discussed in the text are dealt with in the exercises at the end of the chapter.

1   Some Data Features

Before outlining the mechanics of applying specific tests, let us discuss briefly certain data features the analyst is likely to face in practice. Consider the data presented in Table 5.1. Imagine that these are from an experiment in which patients diagnosed as suffering from a certain disease were randomly assigned to one of two treatment regimens, A and B, and were followed up until a certain fixed date or death, whichever came first, unless, of course, they were lost to follow-up earlier. The following features of the data in Table 5.1 are worth noting.

1. Patient a died 8 days after randomization, but before treatment could be started. The prevailing wisdom is that the patient should not be excluded from the analysis (Peto *et al.*, 1977). The argument is this: whereas it is advisable to

Table 5.1

Hypothetical Data from a Clinical Study

Case	Treatment	Date of randomization	Date of death	Trial time
a	A	10/6/1968	Died 18/6/1968[a]	8
b	B	18/10/1970	Died 16/4/1971	180
c	B	9/3/1968	Died 27/4/1974[b]	2,240
d	A	6/6/1969	Died 14/8/1969	220
e	B	4/7/1972	Died 17/7/1972	13
f	B	19/12/1968	Still living[c]	1,989
g	A	19/5/1968	Lost after 4 yrs.	1,460
h	A	10/8/1970	Still living[d]	1,329
i	A	8/2/1970	Out-mig. 8/2/1971	365

[a] Died before treatment could be applied.

[b] Died in an accident.

[c] Changed treatment from B to A on 19/12/1969; still living as of the end of the observation period (31/5/1974)

[d] Still living as of the end of the observation period (31/5/1974).

avoid the occurrence of such situations by delaying randomization until the last possible minute, once the randomization has been done, the analyst should regard it as irrelevant whether the treatment regimen has been completed, if the failure to complete the regimen was due to factors outside the control of the experimenter.

2. Patient c died by accident. The cause of death is extraneous to the experiment. Hence the case is to be treated as though the patient had out-migrated and was lost to follow-up on the date of death (see cases g and i).

3. Patient f changed treatment B to A one year after the date of entry (randomization), and was still alive on the last follow-up date (May 31, 1974). Several options are available, one of them being to proceed as though no change of treatment has occurred. This would mean, of course, that the research question answered is not whether A is better than B for the disease, but whether it is better to adopt a policy of treatment A, with deviations if necessary, or a policy of treatment B, with deviations if necessary. Another option is to exclude the case from the analysis. Yet another option is to treat the case as though the patient had out-migrated and been lost to follow-up. Both these latter options involve sacrificing some information.

4. Patient g was lost to follow-up. The only information available about this patient is that he (she) lived more than $4 \times 365 = 1,460$ days after randomization. This is a censored observation.

5. Patient h was alive on the last follow-up date. In this case all that is known is that the patient lived more than 1,329 days after randomization. This is another censored observation.

6. Patient i out-migrated and was lost to follow-up on February 8, 1971, i.e., 365 days after randomization. Here is another censored case.

7. The last column of Table 5.1 shows the trial times (i.e., the interval between randomization and death or censoring as the case may be). We have used a day as the unit for calculating trial times. One may use a week, month, or year instead. The rule to follow is to use smaller units if the event of interest (e.g., death, failure) occurs very rapidly after the date of entry.

Let us now shift attention to data from surveys or observational studies. Consider the hypothetical data in Table 5.2 showing date of entry, date of termination, and date of interview. Trial times (i.e., the intervals between date of entry and the date of termination or censoring as the case may be) are shown in the last column. Note that in four cases termination had not occurred as of the date of interview. For these (censored) cases we know only that the duration to termination is longer than the trial time shown.

Depending upon the event of interest and the definition of entry time the duration to termination might represent the duration of pospartum amenorrhea after the birth of a child, the interval between marriage and the first pregnancy, the internal between the fitting of the IUD and its subsequent

Table 5.2

Hypothetical Data from a Survey

Case	Subclass	Date of entry	Date of termination[a]	Date of interview	Trial times (days)
a	A	1/1/1979	10/2/1979	24/5/1979	40
b	A	18/1/1979	15/5/1979	18/6/1979	117
c	B	20/1/1979	—	9/5/1979	109+
d	A	30/1/1979	13/2/1979	15/6/1979	14
e	B	28/2/1979	—	23/5/1979	84+
f	B	10/3/1979	8/5/1979	10/5/1979	59
g	A	12/4/1979	—	20/5/1979	30+
h	A	12/5/1979	—	16/5/1979	4+

[a] A dash indicates that termination had not occurred before the interview.

removal or expulsion, the duration of continuous employment under one employer, the waiting time for promotion, etc.

In studies using nations as units of analysis, the date of entry might stand for the birth of each nation and the date of termination for the date of the first coup d'etat therein. If small business organizations are the units of analysis, the date of entry might stand for the date of starting the business and the date of termination the date of going out of business.

A major contrast between survey and observational data on the one hand and experimental data on the other is that subclasses in survey (observational) research are treated as the counterparts of treatment groups in experimental studies. This difference would be immaterial if it were not for the fact that whereas treatments are assigned at random in experiments, subclasses in survey (observational) data are not formed in that manner, and yet the same procedures are used in making subclass comparisons as in making treatment comparisons. Many textbooks extoll the superiority of experiments over nonexperimental research for making causal inferences. This is not the place to examine this controversial topic. For an excellent discussion we refer the reader to Blalock's (1964) *Causal Inference from Non-Experimental Research.*

2   Two-Sample Tests

In this section we outline the application of each of the following nonparametric tests for comparing two treatments or subclasses in terms of their impact on survival functions:

 i. Gehan's generalized Wilcoxon test,

ii. Cox–Mantel test,

iii. the log-rank test, and

iv. Peto and Peto's generalization of the Wilcoxon test.

Theoretical backgrounds of these tests are not discussed, and emphasis in the presentation is on the mechanics of application. The data used in the illustrations are all hypothetical.

Gehan's Generalization of the Wilcoxon Test

Suppose in an experiment 12 communities (local areas) were randomly exposed to one of two family planning programs. Defining survival time as the interval between randomization and the date of attainment of a prespecified target such as sterilization of a certain percentage of ever-married women of reproductive ages, imagine that the following are the survival times observed in the experiment:

Program 1	24, 34, 42, 15+, 40+, 42+
Program 2	10, 25, 27, 34, 41, 12+

the plus sign indicating censored observations. The steps in applying Mantel's (1967) version of Gehan's (1965a,b) generalized Wilcoxon test are shown in Table 5.3.

In large samples the test statistic G (see footnote d to Table 5.3) can be treated as a standard normal variate. For illustrative purposes we so regard the statistic calculated in Table 5.3, and infer, from the comparison of the calculated value ($= 1.417$) with the tabulated 5 percent value ($= 1.96$) of the standard normal variate, that the data do not indicate any difference between the two programs.

It may be noted that the sampling distribution of w (see note b in Table 5.3) can be obtained by considering all possible selections of n_1 U's out of $n_1 + n_2$ U's. In the illustration presented in Table 5.3, however, the large-sample normal approximation of w has been used.

It may also be noted that since w, as defined above, is not continuous, a continuity correction is recommended by some authorities. The correction involves adding 1 to w if there are no ties or censoring and 0.5 otherwise.

The Cox–Mantel Test

Suppose at a university over a period of several years 44 faculty members (20 females and 24 males) were recruited below the rank of full professor, and

Table 5.3

Steps[a,b,c,d] in Applying Gehan's Generalization of the Wilcoxon Test: Mantel's Version

Ordered observations from smallest to largest	Rank uncensored observations from smallest to largest	Assign the next highest rank to the censored case	Reduce rank of tied observation to the lowest rank for the value R_{1i}	Rank observed from smallest to largest	Do step 3 and reduce rank of censored observation 1 R_{2i}	$U_i = R_{1i} - R_{2i}$
	(1)	(2)	(3)	(4)	(5)	(6)
10	1	1	1	12	12	-11
12+		2	2	11	1	1
15+		2	2	10	1	1a
24	2	2	2	9	9	$-7a$
25	3	3	3	8	8	-5
27	4	4	4	7	7	-3
34	5	5	5	6	5	0a
34	6	6	5	5	5	0
40+		7	7	4	1	6a
41	7	7	7	3	3	4
42	8	8	8	2	2	6a
42+		9	9	1	1	8a

[a] a in the last column indicates sample 1.

[b] w = sum of U_i's from sample $1 = 1 - 7 + 0 + 6 + 6 + 8 = 14$

[c] An estimate of the variance of w is

$$v = [n_1 n_2 \Sigma U_i^2]/(n_1 + n_2)(n_1 + n_2 - 1)$$

n_1 and n_2 being the sizes of samples 1 and 2, respectively.

[d] The test statistic is $G = w/\sqrt{v} = 14/\sqrt{97.64} = 1.47$

the accompanying table gives their waiting periods (in years) for promotion to the full Professor rank.

Females: 6, 6, 6, 6+, 8, 9+, 10+, 11, 12+, 13, 16, 17, 18+, 19, 20+, 22, 23, 25+, 32+, 32+
Males: 5, 5, 5, 5, 5, 5, 5, 5, 5+, 5+, 6, 6, 8, 8, 8, 8, 11, 11, 12, 12+, 15, 17, 22+, 23+,

---

A plus sign indicates censoring (which in this case may mean that the faculty member had not been promoted to the full professor rank as of the end of the observation period or that he or she had left the university while at a rank below that of a full professor.)

The hypothesis to be tested is that there has been no sex difference in the waiting time for promotion to the full professor rank. The application of the Cox–Mantel test to the data is shown in Table 5.4.

Column 1 of Table 5.4 shows the distinct promotion times (waiting periods) in the pooled sample. Column 2 shows, for each promotion time, the observed number of promotions among females. Column 3 gives the corresponding information for males, and column 4 for both sexes combined. Column 5 shows the number of females at risk at each promotion time, or more correctly just before each promotion time, the next two columns giving the corresponding information for males and for both sexes combined, respectively. Note that a person is at risk if he or she is in the group in which the event of interest (promotion in the present case) occurs. In calculating the number of persons at risk, the concept of a hypothetical cohort is used. Thus all the 20 females in the sample, irrespective of when they joined the university, are regarded as at risk in the beginning, i.e., at $t = 5$, their number diminishing in accordance with the numbers of promotions and censoring. (Why is R_f the same for $t = 6$ as at $t = 5$? Because there were no promotions or censoring at $t = 5$.)

From the figures in Table 5.4, it is easy to calculate the "expected number" of promotions among persons of each sex under the assumption of no sex difference in the promotion pattern. This is done by distributing the number of promotions at each t in proportion to the numbers of males and females at risk at t (or more correctly just before t). Thus the expected numbers of promotions at $t = 6$ among males and females, respectively, are $3 \times \frac{14}{34}$ and $3 \times \frac{20}{34}$. In symbols then

$$E_f = \sum_t e \frac{R_f}{R} \quad \text{and} \quad E_m = \sum_t e \frac{R_m}{R}$$

where E_f and E_m are the expected numbers of promotions among females and males, respectively, calculated under the assumption of no sex difference in promotion patterns. The observed numbers of promotions are $O_f = 11$ for females and $O_m = 19$ for males (see Table 5.4). The Cox–Mantel test consists

Table 5.4

Cox–Mantel Test Using Statistic C [a,b]

Distinct promotion times (pooled sample)	Number of promotions during t			Number at risk		
	Females e_f	Males e_m	Total e	Females R_f	Males R_m	Total R
(1)	(2)	(3)	(4)	(5)	(6)	(7)
5		8	8	20	24	44
6	3	2	5	20	14	34
8	1	4	5	16	12	28
11	1	2	3	13	8	21
12		1	1	12	6	18
13	1		1	11	4	15
15		1	1	10	4	14
16	1		1	10	3	13
17	1	1	2	9	3	12
19	1		1	7	2	9
22	1		1	5	2	7
23	1		1	4	1	5
Total	11	19	30	—	—	—

[a] Observed number of promotions among males = 19. "Expected" number of promotions among males assuming no sex difference is $\Sigma(e \times R_m/R) = 12.033$.

[b] U = observed − expected = 6.967; estimated variance of U = 6.0819; Cox–Mantel statistic $C = U/\sqrt{6.0819} = 2.825$.

in examining whether O_f is significantly different from E_f (or whether O_m is significantly different from E_m). Writing $U = O_m - E_m$, we test whether U is significantly different from zero. An estimate of the variance of U is obtained by using the following formula:

$$I = \sum_t \frac{e(R - e)R_m R_f}{(R - 1)R^2}$$

or in terms of the columns of Table 5.4:

$$I = \sum \frac{(\text{col. 4})(\text{col. 7} - \text{col. 4})(\text{col. 5})(\text{col. 6})}{(\text{col. 7} - 1)(\text{col. 7})^2}$$

From Table 5.4 we have $U = 6.967$ and $I = 6.0819$. The Cox–Mantel test statistic is (Cox, 1959, 1972; Mantel, 1966)

$$C = \frac{U}{\sqrt{I}}$$

which in large samples is normally distributed with zero mean and unit variance. Since the calculated value of $C (= 2.825)$ has a very small probability level (less than 0.01) we infer that the hypothesis of no sex difference in promotion pattern is not consistent with the data.

The use of the term "expected" number herein may be objected to on the ground that the application of the method of calculating it may lead in some instances to the paradoxical situation with the expected number of events in a group exceeding the number of subjects originally belonging (assigned, randomized) to the group. An alternative term suggested in the literature is "extent of exposure" (Peto et al., 1977). We find this latter term less revealing, and on that ground prefer to use the more familiar, although technically less sound, term "expected number."

The Log-Rank Test

The third test we shall describe is known as the log-rank test (Peto and Peto, 1972). It is based on a set of scores w_i assigned to the various observations. Table 5.5 provides an illustration using the data on faculty promotions referred to earlier.

Column 1 of Table 5.5 gives the promotion and censoring times arranged in ascending order. Column 2 gives the number of persons at risk at each distinct time of promotion. Columns 3 and 4 give the timing of promotions and censorings in the pooled sample and column 5 gives the cumulation of column 3/column 2 starting from the top. The next two columns give the score associated with each promotion and each censoring time. For each distinct promotion time the score is one minus the value in column 5. Thus, for each promotion at $t = 5$, the score is $1 - 0.1818$, i.e., 0.8182. For censored cases at $t +$ the score is the negative of the number for t in column 5. Thus each censored case at $t = 5$ gets a score of -0.1818.

The sum of the w scores is zero for the two samples together. That is the weighted sum of the scores in columns 6 and 7, using the numbers of promotions and censorings as weights, is zero.

The log-rank test is based on the sum S of the w scores for one of the two samples: $S = \Sigma$ col. (6) $\times$ col. (8) $+ \Sigma$ col. (7) $\times$ col. (9). The variance V of S is estimated by using the following formula, where v stands for an estimate of V:

$$v = \frac{n_1 n_2}{(n_1 + n_2)(n_1 + n_2 - 1)} \sum_1^{n_1 + n_2} w_i^2$$

From Table 5.5 we have for males $S = 0.8182 \times 8 - 0.182 \times 2 - \cdots - 1.542 \times 1 = 6.97$ and

$$v = \frac{20 \times 24}{44 \times 43} [(0.8182)^2 \times 8 + (-0.1818)^2 \times 2 + \cdots] = 6.4528$$

Table 5.5

The Log-Rank Test

					Scores		Among males	
$t_{(i)}$	Number at risk	Number of promotions	Number of censorings	$\sum \dfrac{\text{col. (3)}}{\text{col. (2)}}$	Uncensored	Censored	promotions	censorings
(1)	(2)	(3)	(4)	(5)	(6)	(7)	(8)	(9)
5	44	8	2	0.1818	0.8182	−0.1818	8	2
6	34	5	1	0.3289	0.6711	−0.3289	2	
8	28	5	2	0.5074	0.4926	−0.5074	4	
11	21	3	0	0.6503	0.3497	−0.6503	2	
12	18	1	2	0.7059	0.2941	−0.7059	1	1
13	15	1	0	0.7725	0.2275	−0.7725		
15	14	1	0	0.8440	0.1560	−0.8440	1	
16	13	1	0	0.9209	0.0791	−0.9209		
17	12	3	1	1.0875	−0.0875	−1.0875	1	
19	9	1	1	1.1987	−0.1987	−1.1987		
22	7	1	1	1.3415	−0.3415	−1.3415		1
23	5	1	4	1.5415	−0.5415	−1.5415		1

The test statistic for the log-rank test is $L = S/\sqrt{v}$, which is distributed normally with zero mean and unit variance in large samples. From Table 5.5 we have $L = 2.742$, which if treated as a standard normal variate, has a probability level of 0.006. We draw the inference from this that the data are not consistent with the hypothesis of no sex difference in the waiting time for promotion to the full professor rank.

Some remarks are in order about the log-rank test just described.

1. The w_i scores are based on the logarithm of the survivor function. This can be seen by noting that the negative of the figure in column 5 against time t is an estimate of the logarithm of the survival function at t.

2. The numerator of the test statistic L is identical with the numerator of the Cox–Mantel test statistic C, provided that both are calculated for the same sample (e.g., males).

3. The variance V given above is known as the permutation variance, since it is obtained by considering the permutational distribution of $(n_1 + n_2)/n_1!n_2!$ S's.

Peto and Peto's Generalization of the Wilcoxon Test

A fourth method for comparing life tables is Peto and Peto's (1972) generalization of Wilcoxon's two-sample rank sum test. This method consists in using the Kaplan–Meier (1958) PL estimate of the survival function to score the censored and uncensored observations and then proceeding as in the case of the log-rank test.

To illustrate suppose ten students were admitted in a given year to a particular graduate program at a university—a program that allows the students to elect one of two possible tracks (specializations). If the times (in years) taken by the students to complete the program are

Track A	3.5, 5.0+, 6.5, 12.25
Track B	4.25+, 6.5, 8.5+, 10+, 10, 12

what can we say about the difference between the two tracks with respect to the time required for the completion of the program?

Table 5.6 shows the computations involved. Column 3 gives the Kaplan—Meier PL estimate of the survival function and column 4 the w_i scores derived therefrom. Thus for $t = 6.5$, the w score is $0.643 + 0.900 - 1 = 0.543$; for $t = 8.5+$, the score is $0.643 - 1 = -0.357$; for $t = 10$, it is $0.643 + 0.482 - 1 = 0.125$; and so on. Note that as in the log-rank test the w_i scores must sum to zero. The

Peto and Peto test statistic PP is calculated in the same way the L statistic of the log-rank test is calculated. The sum of the w scores for sample B is 0.125. The sum of squares of the scores in the pooled sample is 2.483906, from which we obtain $v = 0.6624$ using the same formula given earlier for the log-rank test: $[4 \times 6/(10 \times 9)]$ times the sum of squares of the w scores.

Relative Merits of the Various Tests

The decision to choose one test over another in a given situation is usually made on the basis of what is known about their powers, i.e., the relative frequency with which each lead to the rejection of false hypotheses. On the basis of a Monte Carlo study, Lee *et al.* (1975) report that when samples are drawn from exponential distributions, with or without censoring, the Cox–Mantel and the log-rank tests are more powerful than Gehan's and Peto and Peto's generalizations of the Wilcoxon test. They also report that in those circumstances there is little difference between the Cox–Mantel test and the log-rank test. One situation identified by them in which the generalized Wilcoxon tests are more powerful than the Cox–Mantel and the log-rank tests is when the ratio of the hazard functions for the two populations involved varies over time. It may be worth noting that in many practical situations the ratio of the hazard functions of two populations is more likely to vary over time (age, duration) than to remain constant. In particular, life (mortality) tables reveal a tendency for the ratio to approach unity with advance to age.

Table 5.6

Peto and Peto's Generalization of Wilcoxon's Test

Completion time t (years)	Rank	Kaplan–Meir PL estimate of $S(t)$	w_i
3.50	1	$\frac{9}{10} = 0.9$	0.900
4.25 +	2		-0.100^b
5.00+	3		-0.100
6.50	4		0.543^b
6.50	5	$\frac{5}{6} \times 0.771 = 0.643^a$	0.543
8.50+	6		-0.357^b
10.00	7	$\frac{3}{4} \times 0.643 = 0.482$	0.125^b
10.00+	8		-0.518^b
12.00	9	$\frac{1}{2} \times 0.482 = 0.241$	-0.277^b
12.25	10	0	-0.759

^{a} We use 0.643 as our estimate of $S(6.50)$.
^{b} From sample B (track B).

3   *k*-Sample Tests

Suppose that we have $k (\geq 2)$ samples of survival times and the problem is to decide whether they belong to the same population of survival times. An example might be whether there is any difference between Catholics, Protestants, and Jews in respect to the timing of the first birth.

The log-rank test and Peto and Peto's generalizations of the Wilcoxon test can be extended to cover *k*-sample problems of the kind just mentioned. Suppose the data on survival times are arranged as shown below:

		Samples		
1	2	$\cdots$		k
t_{i1}	t_{i2}	$\cdots$		t_{ik}

where for sample j, $i = 1, 2, \ldots, n_j$. Some of the t_{ij}'s may represent censoring times. Let $n = n_1 + n_2 + \cdots + n_k$ and let $w_1, w_2, \ldots, w_n$ be the scores assigned to the observations following the procedures described in the preceding section for the log-rank test or the Peto and Peto generalization of the Wilcoxon test. Let S_j be the sum of the scores for the jth sample. To test the hypothesis that the k samples represent the same survival function we calculate

$$X^2 = \left[\sum (S_j^2/n_j) \right] \Big/ s^2$$

where the summation is over samples, and

$$s^2 = \left(\sum w_i^2 \right) \Big/ (n - 1)$$

where the summation is over subjects.

Under the hypothesis of no difference in the survival functions, X^2 has approximately a chi-square distribution with $k - 1$ degrees of freedom (Peto and Peto, 1972).

Note that when $k = 2$, X^2 reduces to the square of the test statistic PP of Peto and Peto's generalization of the Wilcoxon test, for, when $X = 2, S_1^2 = S_2^2$, since the sum of the w scores is zero (in the pooled sample), and hence

$$X^2 = [(S_1^2/n_1) + (S_2^2/n_2)]/S$$

$$= S_1^2/v = S_2^2/v$$

where $v = s^2/[(1/n_1) + (1/n_2)]$.

Table 5.7

Application of Peto and Pike's List[a] to Interval Between Divorce and Remarriage for Four Age-at-Divorce Groups of White Women in the United States: 1975 Current Population Survey

Time since divorce (years)	Age at divorce							
	Under 25 years		25–34 years		35–44 years		45 years or over	
	Number at risk	Number remarried	Number at risk	Number remarried	Number at risk	Number remarried	Number at risk	Number remarried
0	1154.0	268	1247.5	238	601.5	71	248.5	11
1	818.5	222	926.5	177	491.5	38	213.0	9
2	558.0	143	691.5	107	422.5	36	186.0	11
3	385.0	83	538.0	64	361.5	38	159.5	8
4	271.5	53	436.0	50	302.0	16	136.0	3
5	200.5	33	360.0	34	261.0	18	123.5	1
6	159.5	25	306.5	27	221.5	8	113.5	3
7	129.5	14	263.5	21	199.5	9	96.0	1
8	109.5	9	235.0	14	177.0	2	85.5	0
9	95.0	10	213.5	15	164.5	7	78.0	2
10	82.0	7	186.0	16	148.5	8	67.0	0
11	72.5	7	159.0	8	133.0	5	60.0	0
12	62.5	2	143.0	6	118.5	2	54.5	0
13	55.5	4	129.5	6	104.0	2	49.0	2
14	47.0	4	117.5	5	91.0	1	40.5	0
15	40.5	5	105.0	2	81.0	1	34.5	0
16	35.0	2	95.5	5	75.0	0	31.0	1
17	32.0	4	87.0	5	71.0	1	28.5	0
18	27.0	2	80.0	2	66.0	0	25.5	1
19+	25.0	1	72.5	2	60.0	2	20.0	0

[a] The calculation of E_{ij}'s may be illustrated as follows. In the pooled sample the number at risk at time 0 is 3251.5 ($=1154.0 + 1247.5 + 601.5 + 248.5$). The corresponding number of remarriages is 588 ($=268 + 238 + 71 + 11$). Hence we estimate the expected value corresponding to 268 as $(1154.0)(588/3251.5) = 208.69$, that corresponding to 238 as $(1247.5)(588/3251.5) = 225.60$, and so on.

Peto and Pike's Test

Peto and Pike (1973) proposed a k-sample test, which can be described as follows: Suppose that, in the pooled sample, failures occur at times $t_1 < t_2 < \cdots < t_N$. Let n_{ij} be the number of individuals in sample j still at risk just prior to t_i, and let d_{ij} be the number of failures in sample j at time t_i. Let the corresponding numbers in the pooled sample be n_{i+} and d_{i+}, respectively [$n_{i+} = \Sigma_j n_{ij}$, and $d_{i+} = \Sigma_j d_{ij}$.] Under the hypothesis of no difference between samples, the conditional expected value of d_{ij}, given n_{ij}, can be estimated as $E_{ij} = n_{ij}(d_{i+}/n_{i+})$. Let $E_{+j} = \Sigma_i E_{ij}$. Then the test statistic suggested by Peto and Pike is

$$\sum_j (E_{+j} - d_{+j})^2 / E_{+j}$$

which is treated as a chi square with $(k - 1)$ degrees of freedom, k being the number of samples involved. Application of this procedure to the data in Table 5.7 yielded a test statistic of 342.82, which treated as a chi square with three degrees of freedom leads to the rejection of the hypothesis of no difference between subgroups.

4 Post-Stratification

In this section we examine the use of covariates in life table comparisons. The term covariate stands for any factor that can help to account for some of the differences between the survival times of different individuals (groups, nations, organizations). Some examples of covariates are the following: in clinical studies, any information collected on all the subjects before their entry; in survey-data analysis, background information such as race or ethnicity; in observational studies using nations as the unit of analysis, the socioeconomic development status of each nation at the beginning of the observation period; and so on. Information collected after the treatment is underway may be valuable if it is complete, i.e., if it is collected on all individuals who survive a given length of time after their entry. A similar comment applies to observational studies also.

The use of covariates in life table comparisons is illustrated in a worked example in Table 5.8 and 5.9. The structure of Table 5.8 is almost identical to that of Table 5.4 except for the presence of a covariate and the absence of censoring, the latter being a nonessential simplification. Columns 9 and 10 of Table 5.8 give the expected number of events at each t in each class under the assumption of no subclass difference in survivor function and ignoring the information on the covariate. To test whether the observed numbers significantly differ from the expected numbers, we use the statistic

$$X^2 = \sum \frac{(\text{observed} - \text{expected})^2}{\text{expected}}$$

where the summation is over subclasses. With two subclasses, X^2 can be treated as a chi square with one degree of freedom (Peto *et al.*, 1977). From Table 5.8 we have

$$X^2 = \frac{(11 - 6.959)^2}{6.959} + \frac{(14 - 18.041)^2}{18.041}$$

$$= 3.252$$

which when compared with the chi-square distribution with one degree of freedom leads to the inference that there is no subclass difference in survival function.

Now let us poststratify the sample using the information on the covariate z and repeat the calculations of Table 5.8 separately in each stratum. This is done in Table 5.9.

Combining the expected numbers calculated in the two strata, we have for subclass A

$$E_A = 1.579 + 4.234 = 5.813$$

and for subclass B

$$E_B = 5.421 + 13.766 = 19.187.$$

Comparing these with the corresponding observed numbers one gets

$$X^2 = (11 - 5.813)^2/5.813 + (14 - 19.187)^2/19.187 = 6.031$$

which, when referred to the chi-square distribution with one degree of freedom, has a probability level between 0.025 and 0.01, indicating significant subclass difference in survival function.

Thus in this case poststratification on a covariate has helped in detecting a subclass (treatment) difference that escaped detection when the covariate was ignored. This result can be attributed to the fact that stratification produced more homogeneity between the treatment groups taken for comparison.

The procedure illustrated in Table 5.9 can be easily extended to cases involving covariates with more than two categories. The principle to remember is that for each stratum a measure of the treatment difference is obtained and these measures are then summed over strata to obtain an overall measure of treatment difference. Note that there is no new principle involved if

A Two-Sample Comparison

Subject number i	Covariate z	Treatment	Survival time t	Number of events at t	Number at risk A	B	A and B	Expected number of events A	B
(1)	(2)	(3)	(4)	(5)	(6)	(7)	(8)	(9)	(10)
1	1	A	8	1	11	14	25	0.440	0.560
2	2	B	9	1	10	14	24	0.417	0.583
3	2	A	13	1	10	13	23	0.435	0.565
4	2	A	18	1	9	13	22	0.409	0.591
5	2	A	23	1	8	13	21	0.381	0.691
6	2	B	52	1	7	13	20	0.350	0.650
7	2	B	63 ⌉						
8	2	B	63 ⌋	2	7	12	19	0.737	1.263
9	1	A	70	1	7	10	17	0.412	0.588
10	1	A	76	1	6	10	16	0.375	0.625
11	1	A	180	1	5	10	15	0.333	0.667
12	1	A	195	1	4	10	14	0.286	0.714
13	1	A	210	1	3	10	13	0.231	0.769
14	1	A	220	1	2	10	12	0.167	0.833
15	1	B	365	1	1	10	11	0.091	0.909
16	1	B	632	1	1	9	10	0.100	0.900
17	1	B	690	1	1	8	9	0.111	0.889
18	1	B	700	1	1	7	8	0.125	0.875
19	1	B	851	1	1	6	7	0.143	0.857
20	1	B	1296 ⌉						
21	1	B	1296 ⌋	2	1	5	6	0.333	1.667
22	1	B	1470	1	1	3	4	0.250	0.750
23	1	B	1972	1	1	2	3	0.333	0.667
24	1	A	1976	1	1	1	2	0.500	0.500
25	1	B	2240	1	0	1	1	0	1.000
Sum								6.959	18.041

Table 5.9

Analysis of Data in Table 8 Taking into Account Covariate z

Subject number i	Treatment	Survival time t	Number of events at t	Number at risk A	B	A and B	Expected number of events A	B
			Stratum 1 ($z = 1$)					
1	A	8	1	8	10	18	0.444	0.556
9	A	70	1	7	10	17	0.442	0.588
10	A	76	1	6	10	16	0.375	0.625
11	A	180	1	5	10	15	0.333	0.667
12	A	195	1	4	10	14	0.286	0.714
13	A	210	1	3	10	13	0.231	0.769
14	A	220	1	1	10	12	0.167	0.833
15	B	365	1	1	10	11	0.091	0.909
16	B	632	1	1	9	10	0.100	0.900
17	B	690	1	1	8	9	0.111	0.889
18	B	700	1	1	7	8	0.125	0.875
19	B	851	1	1	6	7	0.143	0.857
20	B	1296 ⎤						
21	B	1296 ⎦	2	1	5	6	0.333	1.667
22	B	1470	1	1	3	6	0.250	0.750
23	B	1972	1	1	2	3	0.333	0.667
24	A	1976	1	1	1	2	0.500	0.500
25	B	2240	1	0	1	1	0	1.000
Sum							4.234	13.766
			Stratum 2 ($z = 2$)					
2	B	9	1	3	4	7	0.429	0.571
3	A	13	1	3	3	6	0.500	0.500
4	A	18	1	2	3	5	0.400	0.600
5	A	23	1	1	3	4	0.250	0.750
6	B	52	1	0	3	3	0	1.000
7	B	63 ⎤						
8	B	63 ⎦	2	0	2	2	0	2.000
Sum							1.579	5.421

for stratification purposes the categories obtained by crossing two or more covariates are used.

Sometimes the analyst may be interested in examining the association between a covariate and survival time. Any of the methods outlined in the section on two-sample tests can be used for the purpose if the covariate is a dichotomy and any of the methods outlined above for k-sample tests if it is a polytomy. To examine covariates jointly, one defines "samples" for k-sample tests in terms of the categories obtained by crossing the covariates of interest (e.g., white males, white females, black males, black females).

A covariate may show a strong association with survival time when it is examined in isolation from other covariates, but the association may be considerably weakened when other covariates are introduced in the analysis (Cox, 1972; Breslow, 1975). Also, the association between one covariate, say z_1, and survival time may be strong when a second covariate, say, z_2, is at one level but very weak when the latter is at another level. In such cases, there is said to be an interaction between the two covariates. It should be borne in mind, however, that when more than a few strata are used, each stratum being in such situations considerably smaller than the whole sample, purely random differences between subclasses (of a covariate) will be more marked in each stratum. Such differences often point in opposite directions, thus giving the impression of an interaction whether it is really there or not.

Problems and Complements

1. The following data (reproduced with permission from the American Medical Association and Mayo Foundation) are from Parker et al. (1946). They pertain to the survival experiences of 2418 males with angina pectoris. Survival time in this example equals the interval between diagnosis and death. Notice that the time of interval is one year.

Years since diagnosis	0	1	2	3	4	5	6	7
Number lost to follow-up	0	39	22	23	24	107	133	102
Number of deaths	456	226	152	171	135	125	83	74
Number entering interval	2418	1962	1697	1523	1329	1170	938	722

Years since diagnosis	8	9	10	11	12	13	14	15
Number lost to follow up	68	64	45	53	33	27	23	27
Number of deaths	51	42	43	34	18	9	6	3
Number entering interval	546	427	321	233	146	95	59	30

The corresponding data (Parker et al., 1946) for 555 females are

Years since diagnosis	0	1	2	3	4	5	6	7
Number lost to follow-up	0	8	8	7	7	28	31	32
Number of deaths	82	30	27	22	26	25	20	11
Number entering interval	555	473	435	400	371	338	285	234

Years since diagnosis	8	9	10	11	12	13	14	15
Number lost to follow-up	24	27	22	23	18	9	7	11
Number of deaths	14	13	5	5	5	2	3	3
Number entering interval	191	153	113	86	58	35	24	14

Apply any two-sample test to investigate whether the survival experiences of the two groups differ significantly from one another.

2. The following are hypothetical survival times (in months) of brain tumor patients receiving three different treatments. Investigate whether there are significant differences in the effectiveness of the three treatments. (Censorings are indicated by +.)

Treatment 1	5	7	13	20+	30	35+	
Treatment 2	4	9	15	23	32	33+	
Treatment 3	5	12	21	28	33	33	36+

3. When there is no censoring, any of a number of nonparameteric tests for comparison of independent samples can be used for k-sample comparison of survival times. One such test is the Kruskal–Wallis (1952) H test. Let N be the total number of observations in the pooled sample, and let n_j be the number of observations in sample j. [We assume that all observations are independent of each other.] Let t_{ij} be the ith observation in the jth sample. The null hypothesis to be tested is that all samples have been drawn from the same population. [In clinical trials, the corresponding hypothesis is that all treatments are equally effective.] To compute the Kruskal–Wallis H statistic, we first rank the observations in the pooled sample. Let r_{ij} be the rank of t_{ij}. Let r_{+j} be the sum of the ranks of the observations in sample j. The statistic H is defined as

$$H = \frac{12}{N(N+1)} \sum_{j=1}^{k} \left(\frac{r_{+j}}{n_j}\right) r_{+j} - 3(N+1)$$

If the null hypothesis is true, H has an asymptotic distribution as a chi square with $(k-1)$ degrees of freedom. When there are ties, each observation is assigned the average of the ranks. To correct for ties, the value of H computed in accordance with the formula given above is divided by $1 - \sum_j T_j/(N^3 - N)$, where $T_j = t_j(t_j + 1)(t_j - 1)$, t_j being the number of tied observations in the jth tied group, $j = 1, \ldots, g$, say, subject to the convention that each untied observation constitutes a tied group of size 1. Apply this test to the data given in Exercise 2 above, treating the censored cases as uncensored.

4. The Mantel–Cox test described in the text can be generalized to the k-sample setup as follows. Recall the notation introduced in connection with the discussion of the Peto–Pike test. (Distinct failure times in the pooled sample: $t_1 < \cdots < t_N$; n_{ij} is the number of subjects at risk just prior to t_i in sample j, d_{ij} the number of failures at t_i in sample j.) Define $E_{ij} = d_{i+}(n_{ij}/n_{i+})$, $E_{+j} = \sum_i E_{ij}$, and $x_j = d_{+j} - E_{+j}$. Cox (1972) has shown, that to test for heterogeneity among samples, one may use the statistic

$$B = (x_1 x_2 \cdots x_{k-1}) \mathbf{C}^{-1} (x_1 x_2 \cdots x_{k-1})'$$

where $\mathbf{C}$ is the conditional dispersion (variance–covariance) matrix of the first $(k-1)$ of the x's $[\sum x_j = 0]$. The (j, j') element of $\mathbf{C}$ can be shown to be

$$\sum_i \frac{d_{i+}(n_{i+} - d_{i+})}{n_{i+} - 1} \frac{n_{ij}}{n_{i+}} \left(\delta_{jj'} - \frac{n_{ij'}}{n_{i+}}\right)$$

where $\delta_{jj'} = 1$ if $j = j'$ and 0 otherwise. The statistic B is distributed in large samples as a chi square with $(k-1)$ degrees of freedom. Apply this test to the data in Table 5.6.

5. To compare two population life tables, one ordinarily does not use any formal statistical tests (see, however, Namboodiri et al. 1975, chapter 1). Descriptive features of the two life tables are presented in a comparative fashion; e.g., the plots of the survival functions (l_x/l_0) are displayed on the same graph. The life expectancy at each age can be similarly compared graphically by plotting the corresponding age patterns. Descriptive measures one could compute from the ordinary life table functions for purposes of comparison include median remaining life time at each age, the ratio of death probabilities [e.g., $_nq_x^{(1)}/_nq_x^{(2)}$] at each age, and so on.

6. Sometimes one may be interested in comparing the life table of a subpopulation to that of the parent population. For example, one may wish to compare the life table for a region with that of the country as a whole. In such situations, the subpopulation does not constitute a random sample from the parent population. Any formal test based on the assumption of statistical independence is strictly speaking indefensible. But if the parent population is very large and the subpopulation small, then, for practical purposes, the parent population may be regarded as a standard against which the subpopulation's data may be compared. Elandt-Johnson and Johnson suggest a testing procedure based on median future life time and another based on expected future life time, when data for individuals are available for the subpopulation. For each age the median remaining life time is computed from the standard life table. Then for each individual in the subpopulation, the interval between age at entry (e.g., onset of a disease) and age at failure (e.g., age at death) is calculated, based on which all individuals are classified into those for whom the interval is lower than the corresponding age-specific median remaining lifetime in the standard life table, and those for whom this is not the case. Under the null hypothesis that there is no deviation from the standard, the dichotomous breakdown of the individuals in the subpopulation can be regarded as a binomial with parameters N and $\frac{1}{2}$, N being the size of the subpopulation. [The data may be incomplete, in the sense that some of the members of the subpopulation may be survivors. As long as all survivors have passed their respective medians, the procedure can be applied without any adjustment.]

7. When the data are complete, for those with age at entry x, the average age at death corresponds to $\mathring{e}_x$ of the standard life table. Propose a test statistic to test the hypothesis that the mortality experience of the subpopulation is not different from that depicted in the life table of the parent population. [Assume that the parent population is sufficiently large to permit treating it as a standard.]

Bibliographic Notes

When there is no censoring, standard nonparametric (distribution-free) tests can be used for the comparison of two or more survival distributions. Gibbons (1976) is a useful reference regarding the various nonparametric tests available. Peto and Peto (1972) suggest a general way to generate rank tests with censored data. The log-rank test seems to have been an outgrowth of the works of Mantel and his collaborators (Mantel and Haenszel, 1959; Mantel, 1963, 1966). Related discussions can be found in Thomas (1969), Crowley (1974), and Crowley and Thomas (1975). Kalbfleisch and Prentice (1980) give a generalization of the Wilcoxon test that is slightly different from Peto and Peto's generalization. The difference is due to the use of different estimators for the survival function. For a review of the use of rank procedures for comparing several survival curves, reference may be made to Ware and Byar (1979). A nontechnical exposition of the log-rank test is available in Peto et al. (1977).

Chapter 6 | Multiple-Decrement Life Tables I

We have considered thus far life tables that describe the attrition caused by a given single factor (e.g., mortality). We now turn to multiple decrements, that is, two or more types of attrition operating together. A simple example of this latter type of life table is one that shows how the single (never married) population diminishes in size over time through mortality and marriage. Another example is a life table that portrays how diminution due to mortality takes place through various causes of death.

Of particular concern to those who have developed multiple-decrement life table methods has been the separate effects of attrition factors. Thus, in cause-of-death analyses using multiple-decrement life tables, one might be interested in estimating the probability that a person will eventually die from a particular cause or in the increase in the expectation of life resulting from the elimination of a particular cause of death.

The basic sources of data for the construction of multiple-decrement life tables are registration statistics and censuses on the one hand and follow-up studies on the other. Data on attrition of singles through mortality and marriage, for example, can be obtained, in many countries, from registration records and census tabulations. Also, cause of death statistics can be obtained from registration sources. If, however, one wants to study the attrition of contraceptors because of pregnancy and for medical and like reasons, the required data may have to be collected via follow-up studies or their equivalents.

The construction of multiple-decrement life tables from follow-up data will be discussed in the next chapter. The present one is concerned with the corresponding exercise using registration statistics and census records. More specifically, the concern of this chapter is with cause-of-death life tables. The emphasis is on computation and interpretation. The mathematical basis of the procedures is discussed in Chapter 8.

1 Computation of Cause-of-Death Life Tables

The steps involved in the construction of multiple decrement life tables dealing with causes of death are:

1. Compute age-specific and age-and-cause specific death rates and cause-of-death ratios.
2. Construct an ordinary life table using the age-specific death rates (for all causes combined).
3. Distribute by cause the total number of deaths in each age group in the life table population obtained in Step 2.
4. Compute the probabilities of eventual death by each specified cause.
5. Compute the probabilities of death by cause in broad age groups, as needed.
6. Compute and plot estimated life tables by cause elimination.
7. Construct associated decrement life tables by cause elimination.

These steps will now be described in detail. For illustrative purposes, we use data from the United States for 1960.

Age-Specific and Age-and-Cause Specific Death Rates

To compute age-and-cause specific death rates we need the numbers of deaths from each cause in each age interval during the year in question and the mid-year base population classified by age. If $_nD_{x,\alpha}$ denotes the (observed) number of deaths in the age group $[x, x + n)$ due to cause C_α in a given year, and $_nN_x$ stands for the corresponding mid-year base population, the age-and-cause specific death rate $_nM_{x,\alpha}$ for the age group $[x, x + n)$ and cause α is given by

$$_nM_{x,\alpha} = (_nD_{x,\alpha}/_nN_x)_\kappa \tag{6.1}$$

where κ is an arbitrary constant, usually 100,000. Thus from Table 6.1, we have for $[65, 70)$ for cardiovasculorenal (CVR) diseases $_5D_{65,\text{CVR}} = 117{,}795$ and $_5N_{65,\text{CVR}} = 6{,}186{,}763$, giving

$$_5M_{65,\text{CVR}} = \frac{117{,}795}{6{,}186{,}763} \times 100{,}000 = 1903.98$$

Table 6.1

Deaths by Cause and Mid-Year Population, U.S. 1960

| Age interval | Mid-year population | Number of deaths by cause | | | | Number of deaths from all causes |
		CVR	Malignant Neoplasm	Accident	Other	
(1)	(2)	(3)	(4)	(5)	(6)	(7)
0-1	4,126,560	501	298	3,835	106,300	110,934
1-5	16,195,304	398	1,762	5,124	10,409	17,693
5-10	18,659,141	401	1,391	3,690	3,686	9,168
10-15	16,815,965	541	1,024	3,152	2,661	7,378
15-20	13,287,434	825	1,023	6,709	3,635	12,192
20-25	10,803,165	1,198	977	6,759	4,422	13,356
25-30	10,870,165	1,973	1,600	5,030	5,620	14,223
30-35	11,951,709	3,912	2,842	4,781	7,675	19,210
35-40	12,508,316	8,350	5,374	4,960	10,452	29,176
40-45	11,567,216	15,939	9,003	4,907	13,112	42,961
45-50	10,528,878	27,566	14,732	5,148	16,864	64,310
50-55	9,696,502	43,121	21,540	5,063	20,906	90,630
55-60	8,595,947	60,750	27,638	4,690	23,681	116,799
60-70	7,111,897	86,238	34,172	4,502	28,592	153,504
65-70	6,186,763	117,795	39,721	4,742	34,424	156,682
70-75	4,661,136	143,339	38,802	4,956	36,699	223,796
75-80	2,977,347	150,378	31,548	5,026	33,113	220,065
80-85	1,518.206	133,473	20,703	4,922	26,207	185,305
85-90	648,581	90,109	10,110	3,607	16,588	120,414
90-95	170,653	38,294	2,839	1,733	7,433	50,299
95+	44,551	10,527	528	470	2,362	13,887
Total	179,325,436	935,668	267,627	93,806	414,881	1,711,982

The age-specific death rate (ignoring causes, i.e., for all causes combined), for the same age group can be computed as the sum of the corresponding age-and-cause specific death rates or more directly as κ times the ratio of the number of deaths (for all causes combined) in the age group in question to the corresponding base population. Thus if we use the notation $_nD_{x+}$ for the sum of $_nD_{x,\alpha}$ over α, then

$$_nM_{x+} = (_nD_{x+}/_nN_x)\kappa = \sum {}_nM_{x,\alpha} \qquad (6.2)$$

where $_nM_{x+}$ stands for the age-specific death rate (for all causes combined).

The age-and-cause specific death rates calculated from the data in Table 6.1 are shown in Table 6.2a along with the age-specific death rates (for all causes combined).

Cause-of-Death Ratios

By expressing the number of deaths from cause C_α in a given age group as a fraction of the total number of deaths in that age group, we obtain the cause-of-death ratio for cause C_α for that age group. For $[x, x + n)$ the cause-of-death ratios are $_nD_{x,\alpha}/_nD_{x,+}$, for the various causes of death (C_α). Obviously, they add up to unity for each age group, provided that causes of death considered are mutually exclusive and jointly exhaustive. In Table 6.1 for the age group $[65, 70)$ we notice 117,795 deaths from CVR, 39,721 from cancer (a short term

Table 6.2

a. Age-and-cause Specific death rates: U.S. population, 1960

Age $[x, x + n)$	Observed age-specific death rates	Age-and-cause specific death rates (per 100,000)			
		CVR	Cancer	Accidents	Other
0–1	0.026883	121	72	929	25760
1–5	0.001092	25	109	316	643
5–10	0.000491	21	75	198	198
10–15	0.000439	32	61	187	158
15–20	0.000918	62	77	505	274
20–25	0.001236	111	90	626	409
25–30	0.001308	182	147	463	517
30–35	0.001607	327	238	400	642
35–40	0.002333	668	430	397	839
40–45	0.003714	1378	778	424	1134
45–50	0.005884	2522	1348	471	1541
50–55	0.009347	4447	2221	522	2156
55–60	0.013588	7067	3215	546	2755
60–65	0.021584	12126	4805	633	4020
65–70	0.031791	19040	6420	766	5564
70–75	0.048013	30752	8326	1063	7873
75–80	0.073991	50507	10596	1688	11122
80–85	0.122055	87915	13636	3242	17262
85–90	0.185658	138933	15588	5561	25576
90–95	0.294744	224397	16636	10155	43556
95+	0.311710	236291	11852	10550	53018

Table 6.2

b. Cause-of-death ratios: U.S. population, 1960

Age	Cause of death			
$[x, x + n)$	CVR	Cancer	Accidents	Other
0–1	0.004516	0.002628	0.034570	0.958227
1–5	0.022495	0.099597	0.289606	0.588312
5–10	0.043739	0.151723	0.402487	0.402051
10–15	0.073326	0.138791	0.427216	0.360667
15–20	0.067667	0.083907	0.550279	0.298146
20–25	0.089698	0.073151	0.506065	0.331087
25–30	0.138719	0.112494	0.353653	0.395135
30–35	0.203644	0.147944	0.248881	0.399531
35–40	0.286194	0.184192	0.170003	0.359611
40–45	0.371011	0.209562	0.114220	0.305207
45–50	0.428643	0.229078	0.080050	0.262230
50–55	0.475792	0.237670	0.055865	0.230674
55–60	0.520467	0.236629	0.040154	0.202750
60–65	0.561796	0.222613	0.029328	0.186262
65–70	0.598911	0.211955	0.024110	0.175024
70–75	0.640490	0.173381	0.022145	0.163984
75–80	0.683334	0.143358	0.022839	0.150469
80–85	0.720288	0.111724	0.026562	0.141426
85–90	0.748327	0.083960	0.029955	0.137758
90–95	0.761327	0.056442	0.034454	0.147776
95+	0.758047	0.038021	0.033845	0.170087

for malignant neoplasm), 4,742 from accidents, and 34,424 from other causes. Their total is 196,682. Expressing the components of this sum as fractions, we get the following cause-of-death ratios for $[65, 70)$:

CVR	$\dfrac{117,795}{196,682} = 0.598911$
Cancer	$\dfrac{39,721}{196,682} = 0.201955$
Accidents	$\dfrac{4,742}{196,682} = 0.024110$
Other	$\dfrac{34,424}{196,682} = 0.175024$

The cause-of-death ratios thus computed are shown in Table 6.2b.

Construction of Ordinary Life Table

Using the age-specific death rates shown in Table 6.2a, an ordinary life table can be constructed following the usual procedures. The life table thus constructed is presented in Table 6.3a, showing $_nq_x$, l_x, and $_nd_x$ values.

Distribution of Life Table Deaths by Cause of Death

Two principal methods are commonly used for distributing the life table deaths by cause of death. One, which may be called the actuarial method, is predicated on the assumption that the cause-of-death ratios in the life table population are the same as the "observed" cause-of-death ratios (computed from the cause-of-death data for the actual population). The following formula

Table 6.3

a. Ordinary life table and the distribution of life table deaths by cause of death

Age [x, x + n)	$_nq_x$	l_x	$_nd_x$	Distribution of life table death by cause			
				CVR	Cancer	Accidents	Other
0–1	0.026525	100000	2652	12	7	92	2542
1–5	0.004360	97348	424	10	42	123	250
5–10	0.002454	96923	238	10	36	96	96
10–15	0.002191	96685	212	16	29	91	76
15–20	0.004577	96473	442	30	37	243	132
20–25	0.006162	96032	592	53	43	299	196
25–30	0.006521	95440	622	86	70	220	246
30–35	0.008004	94818	759	155	112	189	303
35–45	0.011595	94059	1091	312	201	185	392
40–45	0.018399	92968	1711	635	358	195	522
45–50	0.028993	91258	2646	1134	606	212	694
50–55	0.045658	88612	4046	1925	962	226	933
55–60	0.065682	84566	5554	2891	1314	223	1126
60–65	0.102301	79011	8083	4541	1799	237	1506
65–70	0.146964	70928	10424	6243	2105	251	1824
70–75	0.213424	60505	12913	8271	2239	286	2118
75–80	0.308966	47591	14704	10048	2108	336	2213
80–85	0.456799	32887	15023	10821	1678	399	2125
85–90	0.604770	17864	10804	8085	907	324	1488
90–95	0.770929	7061	5443	4144	307	188	804
95+	1.000000	1617	1617	1226	61	55	275
Total				60648	15021	4470	19861

Table 6.3

b. Death probabilities by cause

Age $[x, x + n)$	CVR	Cancer	Accidents	Other
0–1	0.000120	0.000070	0.000920	0.025420
1–5	0.000103	0.000431	0.001264	0.002568
5–10	0.000103	0.000371	0.000990	0.000950
10–15	0.000165	0.000300	0.000941	0.000786
15–20	0.000311	0.000384	0.002519	0.001368
20–25	0.000552	0.000448	0.003114	0.002041
25–30	0.000901	0.000733	0.002305	0.002578
30–35	0.001635	0.001181	0.001993	0.003195
35–40	0.003317	0.002137	0.001967	0.004168
40–45	0.006830	0.003851	0.002097	0.005615
45–50	0.012426	0.006641	0.002323	0.007605
50–55	0.021724	0.010856	0.002550	0.010529
55–60	0.034186	0.015538	0.002637	0.013315
60–65	0.057473	0.022769	0.003000	0.019061
65–70	0.088018	0.029678	0.003539	0.025716
70–75	0.136701	0.037006	0.004727	0.035006
75–80	0.211131	0.044294	0.007060	0.046500
80–85	0.329033	0.051023	0.012132	0.064615
85–90	0.452576	0.050771	0.018137	0.083294
90–95	0.586924	0.043481	0.026627	0.113872
95+	0.758201	0.037716	0.034006	0.170029

(Spiegelman, 1968, p. 137) reflects this approach:

$$_nd_{x,\alpha} = {_nd_{x,+}}({_nD_{x,\alpha}}/{_nD_{x+}}) \tag{6.3}$$

where $_nd_{x,\alpha}$ stands for the number of (life table) deaths at age $[x, x + n)$, attributable to cause C_α; $_nd_{x,+}$ is the total number of (life table) deaths at age $[x, x + n)$; and $_nD_{x,\alpha}$ and $_nD_{x,+}$ are the corresponding observed quantities. Applying this formula to the figures in Tables 6.2a and 6.2b for $[65, 70)$ gives $_5d_{65,\text{CVR}} = 10,424 \times 0.598911 = 6,243$, $_5d_{65,\text{cancer}} = 10,424 \times 0.201955 = 2,105$, $_5d_{65,\text{accident}} = 10,424 \times 0.024110 = 251$, and $_5d_{65,\text{other}} = 10,424 \times 0.175024 = 1,824$. Another method for distributing the total number of deaths in the life table population by cause of death has been suggested by Keyfitz and Frauenthal (1975). This involves applying the following multiplier to the estimate obtained by formula (6.3) given above:

$$1 + \frac{1}{48}\left[\frac{_nP_{x+n} - {_nP_{x-n}}}{_nP_x} + 2n\,_nM_{x,+}\right]$$

$$\times \left[\frac{_nM_{x+n,+} - {_nM_{x-n,+}}}{_nM_{x,+}} - \frac{_nM_{x+n,\alpha} - {_nM_{x-n,\alpha}}}{_nM_{x,\alpha}}\right] \tag{6.4}$$

Table 6.3

c. Expected number of individuals of age x who eventually die of cause C_x: U.S., 1960

Age $[x, x + n)$	Total	CVR	Cancer	Accidents	Other
0–1	100000	60648	15021	4470	19861
1–5	97348	60636	15014	4378	17320
5–10	96923	60626	14972	4255	17070
10–15	96685	60616	14936	4159	16974
15–20	96473	60600	14907	4068	16898
20–25	96032	60570	14870	3825	16767
25–30	95440	60517	14827	3526	16570
30–35	94817	60431	14757	3306	16323
35–40	94058	60276	14645	3117	16020
40–45	92968	59964	14444	2938	15628
45–50	91257	59329	14086	2737	15105
50–55	88611	58195	13480	2525	14411
55–60	84566	56270	12518	2299	13479
60–65	79011	53379	11204	2076	12352
65–70	70928	48838	9405	1839	10846
70–75	60504	42595	7300	1588	9021
75–80	47591	34324	5051	1302	6904
80–85	32887	24276	2953	966	4692
85–90	17964	13455	1275	567	2667
90–95	7061	5370	368	243	1080
95+	1617	1226	61	55	275

where all symbols except $_nP_x$, which denotes the mid-year population in the age group $[x, x + n)$, have already been introduced. From Tables 6.1 and 6.2 we have

x	$_5P_x$	$_5M_{x,+}$	$_5M_{x,\mathrm{CVR}}$
60	7,111,897	0.021584	0.012126
65	6,186,763	0.031791	0.019040
70	4,661,136	0.048013	0.030752

Also recall that the cause-of-death ratio for CVR for [65, 70) is 0.59891. From these figures we obtain the refined estimate of $_5d_{65,\mathrm{CVR}}$ as 6,106, which differs only slightly from the estimate obtained earlier (6,243), using (6.3). As Keyfitz and Frauenthal (1975) point out, when, within an age group, one cause is declining while another is increasing, the refinement proposed by them is likely to make a significant difference, whereas in other situations the refinement

Table 6.3

d. Conditional probabilities of eventually dying from specified causes,
for those alive at age U.S. 1960

Age $(x, x + n)$	CVR	Cancer	Accidents	Other
0–1	0.6065	0.1502	0.0447	0.1986
1–5	0.6229	0.1542	0.0450	0.1779
5–10	0.6255	0.1545	0.0439	0.1761
10–15	0.6269	0.1545	0.0430	0.1756
15–20	0.6282	0.1545	0.0422	0.1752
20–25	0.6307	0.1548	0.0398	0.1746
25–30	0.6341	0.1553	0.0369	0.1736
30–35	0.6373	0.1556	0.0349	0.1722
35–40	0.6408	0.1557	0.0331	0.1703
40–45	0.6450	0.1554	0.0315	0.1681
45–50	0.6501	0.1544	0.0300	0.1655
50–55	0.6557	0.1521	0.0285	0.1626
55–60	0.6654	0.1480	0.0272	0.1594
60–65	0.6756	0.1418	0.0263	0.1563
65–70	0.6886	0.1326	0.0259	0.1529
70–75	0.7040	0.1207	0.0262	0.1491
75–80	0.7212	0.1063	0.0274	0.1451
80–85	0.7382	0.0898	0.0294	0.1427
85–90	0.7490	0.0710	0.0316	0.1485
90–05	0.7605	0.0521	0.0344	0.1530
95+	0.7582	0.0377	0.0340	0.1701

may not be necessary. Columns 5–8 of Table 6.3a show the values of $_nd_{x,\alpha}$ computed in accordance with (6.3).

Cause-Specific Probability of Dying

By dividing $_nd_{x,\alpha}$ by the corresponding l_x, we get the probability that an individual will die of cause C_α after surviving to age x but before reaching age $x + n$, when all causes are effective. These are often called the *crude probabilities of death* from specified causes. The numbers thus calculated are shown in Table 6.3b. Thus the crude probability of dying from CVR in the age group [65, 70) is estimated as $\frac{6,243}{70,928} = 0.088018$.

Conditional Probability of Eventually Dying from a Given Cause

If we obtain the sum of $_nd_{x,\alpha}$ for age groups [65, 70) and up, and divide the sum by l_x we get the probability that an individual who has survived to age 65

Table 6.3

e. Densities of cause of death in the presence of all causes

Age [x, x + n)	Midpoint	CVR	Cancer	Accidents	Other
0–1	0.5	0.000198	0.000466	0.020582	0.127939
1–5	3.0	0.000041	0.000699	0.006879	0.003147
5–10	7.5	0.000033	0.000479	0.004295	0.000967
10–15	12.5	0.000053	0.000386	0.004072	0.000765
15–20	17.5	0.000099	0.000493	0.010872	0.001319
20–25	22.5	0.000175	0.000573	0.013378	0.001984
25–30	27.5	0.000284	0.000932	0.009843	0.002487
30–35	32.5	0.000511	0.001491	0.008456	0.003051
35–40	37.5	0.001029	0.002676	0.008277	0.003947
40–45	42.5	0.002094	0.004767	0.008725	0.005267
45–50	47.5	0.003740	0.008069	0.009485	0.006989
50–55	52.5	0.006348	0.012809	0.010872	0.009285
55–59	57.5	0.009534	0.017496	0.013378	0.011349
60–65	62.5	0.014975	0.023953	0.009843	0.015165
65–70	67.5	0.020588	0.028027	0.008456	0.018378
70–75	72.5	0.027275	0.029812	0.008277	0.021318
75–80	77.5	0.033135	0.028067	0.008725	0.022275
80–85	82.5	0.035685	0.022342	0.009485	0.020392
85–89	87.5	0.026662	0.012076	0.010112	0.015981
90–95	92.5	0.013666	0.004088	0.009978	0.008106
95+		—	—	—	—

will eventually die of cause C_α. To illustrate from Table 6.3a, we have for CVR the sum of the figures in the death column beginning with age group [65, 70) and going up, equal to $6,243 + \cdots + 1,226 = 48,838$, which when divided by $l_{65} = 70,928$, gives 0.68856 as our estimate of the probability that an individual who has survived to age 65 will eventually die of CVR when all causes are in effect. The figures thus computed are presented in of Table 6.3d. Notice that according to these figures the probability that a newborn baby will die of CVR is 0.6065. A rising pattern by age until age 95 is clearly seen for these probabilities. The probability of eventually dying of cancer, on the other hand, increases with age, reaches a maximum by about 40, and then declines. The corresponding pattern for accidents is clearly a declining one throughout life except in infancy.

Probability of Dying from a Given Cause Conditional on Dying in a Given Age Interval

From the l_x column we can compute, by differencing, the number of persons dying between two birthdays, e.g., between the 5th and the 40th. The same can be computed from the figures in the total death column also. The corresponding number of deaths due to a given cause can be similarly computed from the death column for that cause. Dividing the latter by the former gives the probability that an individual will die of the particular cause within the age interval, given that the individual dies within that age interval. Thus for the age range [5, 35) the probabilities of dying of CVR, cancer, accidents, and other causes are

$$(60{,}626 - 60{,}276)/(96{,}923 - 94{,}058) = 0.1222$$

$$(14{,}972 - 14{,}645)/(96{,}923 - 94{,}058) = 0.1141$$

$$(4{,}225 - 3{,}117)/(96{,}923 - 94{,}058) = 0.3972$$

and

$$(17{,}070 - 16{,}020)/(96{,}923 - 94{,}058) = 0.3665$$

respectively. These figures show that for persons in the age range [5, 35), the most likely cause of death is accident. Similarly, we find that for persons of age 45, the most likely cause of death over the age range [45, 75) is CVR.

In general, if we define $l_{x,\alpha}$ as the sum of all (life table) deaths from cause C_α that occur in the age range $[x, \infty)$, that is,

$$l_{x,\alpha} = {}_n d_{x,\alpha} + {}_n d_{x+n,\alpha} + \cdots$$

The summation being carried to the end of the life table, then

$$(l_{x,\alpha} - l_{x+a,\alpha})/(l_x - l_{x+a})$$

for $a > 0$ gives the probability that a person who dies in the age range $[x, x + a)$ dies of cause C_α.

Cause-Specific Density Function

The graph obtained by plotting

$$(l_{x,\alpha} - l_{x+n,\alpha})/n l_{0,\alpha}$$

against the mid-point of $[x, x + n)$ is sometimes called the curve of death from cause C_α. These are approximate density functions evaluated at midpoints $x + \frac{1}{2}n$. From Table 6.3a, we have $l_{65,\text{CVR}} = 48{,}838$, $l_{70,\text{CVR}} = 42{,}595$, and $l_{0,\text{CVR}} = 60{,}648$. These figures give the approximate density function evaluated at 67.5 as $(48{,}838 - 42{,}595)/[5 \times 60{,}648] = 0.0206$. The quantities thus calculated for each of the four causes are shown in Table 6.3e. For purposes of these calculations the last age interval was closed (arbitrarily) by assuming that no one survives beyond 110. The curves of death for the four causes are shown in Fig. 6.1. Notice that the curve for accidents is

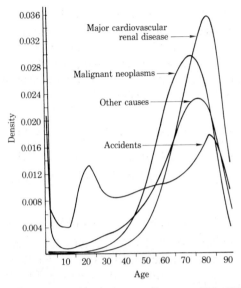

Fig. 6.1. Curves of death, four specific causes, U.S., 1960.

different from the others in that it shows a tendency to be trimodal, with peaks at the beginning, in the age interval [20, 25), and in the age interval [80, 85). The modal age for CVR is between 80 and 85, that for cancer is between 70 and 75, and for the residual ("other") causes, it is between 75 and 80.

2 Cause Elimination

When infectious diseases were the main killers, it was natural to think of eliminating one or more causes of death by eradicating the sources of infection. Thus William Farr (1874) found it worthwhile to examine the "Effect of the extinction of any single disease on the duration of life" in a supplement to the *Annual Report of the Registrar General* (England). In that report Farr refers to earlier discussions on the topic by Daniel Bernoulli and d'Alembert. More modern developments do not bear much resemblance to the very early works on the subject (see Birnbaum, 1979, for a brief sketch of the historical development of the literature). Nowadays attention seems to be focused on variously defined net probabilities of survival until a given age or of dying in a given age interval, given survival to the beginning of that interval. One type of net probability is the probability that a person alive at the beginning of an age interval $[x, x + n)$ will die before reaching the end of the interval, if the population is exposed to the risk of dying from a cause, say C_α, only. Another type net probability is the probability that a person alive at the beginning of an age interval $[x, x + n)$ will die before reaching the end of that interval if the risk of dying of cause C_α were "eliminated" in some sense. Such net probabilities can be estimated using any of a number of formulas suggested in the literature. Some of these are mentioned in the problems and complements at the end of the Chapter. We note here that, ordinarily, the estimates obtained using any of the various formulas do not differ markedly from those obtained using any other. The reasoning behind one of these formulas is outlined below.

Suppose we are interested in the effect on survival probability of "removing" all but one cause of death. *We assume that the diverse causes act independently of each other*, which is the same as saying that their respective forces of mortality are additive; that is,

$$\mu(x) = \mu_\alpha(x) + \mu_{(-\alpha)}(x)$$

where $\mu_\alpha(x)$ is the force of mortality due to cause C_α, and $\mu_{(-\alpha)}(x)$ is the force of mortality due to all other causes combined. If the survival probability when the only effective cause of death is C_α is $\bar{l}_\alpha(x)$, we have from the additivity of forces of mortality assumed above,

$$l_x = \bar{l}_\alpha(x)\bar{l}_{(-\alpha)}(x)$$

Now applying the relationship between survival function and force of mortality, we note that

$$-\ln \frac{\bar{l}_\alpha(x + n)}{\bar{l}_\alpha(x)} = \int_0^n \mu_\alpha(x + t)\, dt$$

$$= \left[\int_0^n \mu_\alpha(x + t)\, dt \bigg/ \int_0^n \mu(x + t)\, dt \right] \int_0^n \mu(x + t)\, dt$$

$$= R \int_0^n \mu(x + t)\, dt,$$

say. It follows that

$$\bar{l}_\alpha(x + n)/\bar{l}_\alpha(x) = [l(x + n)/l(x)]^R$$

This implies that once the ordinary life table for all causes combined is on hand, we can compute the ratios $\bar{l}_\alpha(x + n)/\bar{l}_\alpha(x)$ by simply raising the corresponding ratio $l(x + n)/l(x)$ to the appropriate power R. If we assume that the force of mortality in a given age interval can be approximated by the corresponding age-specific death rates, then it follows that

$$R = \int_0^n \mu_\alpha(x + t)\, dt \bigg/ \int_0^n \mu(x + t)\, dt$$

$$= {}_nM_{x,\alpha}/{}_nM_{x,+} = {}_nD_{x,\alpha}/{}_nD_{x,+}$$

Table 6.4

Effect of Elimination of CVR Deaths

Age $[x, x + n)$	Death probabilities			Expectation of life		
	All causes	CVR only	All but CVR	All causes	CVR only	All but CVR
65–70	0.146964	0.090808	0.061765	14.11	17.14	26.72
70–75	0.213424	0.142527	0.082681	11.12	13.60	23.31
75–80	0.308966	0.223176	0.110436	8.49	10.46	20.19
80–85	0.456799	0.355695	0.156920	6.23	7.78	17.39
85–90	0.604770	0.500765	0.208329	4.58	5.79	15.18
90–95	0.770929	0.674364	0.296542	3.35	4.38	13.54
95+	1.000000	1.000000	1.000000	3.21	4.23	13.26

(see Greville, 1948; Chiang, 1968). Net probabilities calculated in this fashion for CVR from the 1960 U. S. data introduced earlier are presented in Table 6.4 for ages 65 and over. The corresponding life expectancies are also shown in the table. A comparison of the life expectancies thus calculated with those of the life table for all causes combined tells us how severe the cause CVR seems to be. The gain in life expectancy at age 65 attributable to eliminating CVR is $26.72 - 14.11 = 12.61$ years. The corresponding figure for all other causes combined is $17.14 - 14.11 = 3.03$ only. The relative magnitudes of these figures is indicative of the major impact CVR has on the mortality experiences of persons 65 years and older.

Problems and Complements

1. Some authors use the notation al_x, aq_x, ad_x, etc., for the functions of the life table when all causes are combined (the prefix a indicating the presence of *all* causes). See, for example, Elandt–Johnson and Johnson (1980). In *English* life tables the corresponding notation is $(al)_x$, $(aq)_x$, $(ad)_x$, and so on. The prefix n to indicate the length of the age interval is also common, e.g., $_n(aq)_x$, or $_n aq_x$, $_n(ad)_x$, or $_n ad_x$. For cause C_α, one finds in English life tables the notation $_n(aq)_x^\alpha$, $_n(ad)_x^\alpha$, and so on. We have not followed this practice in our exposition. When the reference is to all causes combined, we have made that explicit by stating it verbally when necessary.

2. As mentioned in the text the net probabilities can be computed using any of a number of formulas. One simple formula is the following:

$$_n q_{x,\alpha} = {}_n d_{x,\alpha} / \{l_x - \tfrac{1}{2}{}_n d_{x,(-\alpha)}\}$$

where $(-\alpha)$ refers to all causes except C_α. Note that l_x is from the life table for all causes combined, and the denominator on the right-hand side represents the number of persons in the life table population surviving until age x, less one-half of $[_nd_x - {_nd_{x,a}}]$. This formula treats deaths from causes other than C_a as withdrawals.

3. Another formula that may be used for computing the net probabilities is the following:

$$_nq_{x,a} = 1 - \exp[-n_nM_{x,a}]$$

What is the rationale behind this formula?

4. Compute $_5q_{x,\text{CVR}}$, from the data presented in the text using the various formulas. Ans. The formula in Exercise 2 above gives 0.090691; the one in Exercise 3 gives 0.090809; the one presented in the text gives 0.090888. More often than not the various formulas give approximately the same results.

5. Estimate $_5q_{65,\text{CVR}}$ from the data presented in the text using the following formula suggested by Keyfitz and Frauenthal (1975)

$$l_{x+n,a}/l_{x,a} = (l_{x+n}/l_x)R$$

Table 6.5

Computation of Keyfitz's H_a for CVR from U.S. Data, 1960

Age	$_nM_{x,\text{CVR}}$ [a]	$_nM_{x,+}$ [b]	$l_{x,\text{CVR}}$ [c]	l_x [d]	$l_{x,\text{CVR}}l_x\varphi_x$ [e,f]
(1)	(2)	(3)	(4)	(5)	(6)
[65, 70)	0.019040	0.031791	1.00000	1.00000	4.42
[70, 75)	0.030752	0.048013	0.90920	0.85304	3.21
[75, 80)	0.050507	0.073913	0.77960	0.67098	1.95
[80, 85)	0.087915	0.122055	0.60562	0.46366	0.87
[85, 90)	0.138933	0.185658	0.39020	0.25186	0.24
[90, 95)	0.224397	0.294744	0.19480	0.09954	0.03
[95, $)	0.236291	0.311710	0.06344	0.02280	0.00

[a] From Table 6.2a for CVR.

[b] From Table 6.2a for all causes combined.

[c] Based on the $q_{x,\text{CVR}}$ values in Table 6.4. These represent the ordinates of the survival function for those alive at age 65 and are subjected only to CVR deaths.

[d] Based on the $q_{x\cdot\text{CVR}}$ values shown in Table 6.4. These figures represent the survival function for those alive at age 65 and are subjected to the risk of dying from any cause other than CVR.

[e] Products of three quantities: (col. 4) × (col. 5) × φ, where φ has been calculated in accordance with the formula

$$\varphi = \{1 - \exp[-5(\text{col. } 2 + \text{col. } 3)]\}/(\text{col. } 2 + \text{col. } 3)$$

Thus $4.42 = 1.0 \times 1.0 \times [1 - \exp(-0.254155)]/0.050831$. For the last age group $\varphi = 1/(\text{col. } 2 + \text{col. } 3)$.

[f] The sum of the numbers in column 6 is 10.72. From Table 6.4 in the text we note that life expectancy at age 65 when all causes are effective is 14.11. The value of $H_{65,\text{CVR}}$ is thus $1 - (10.72/14.11) = 0.24$. Thus a uniform reduction in the force of mortality due to CVR at all ages 65 and over by one percent would result in a 0.24 percent increase in the life expectancy at age 65.

where

$$R = \frac{{}_nM_{x,\alpha}}{{}_nM_x}\left\{1 + \left(\frac{1}{24}\right)\left[\frac{{}_nM_{x+n,\alpha} + {}_nM_{x-n,\alpha}}{{}_nM_{x,\alpha}} - \frac{{}_nM_{x+n} + {}_nM_{x-n}}{{}_nM_x}\right]\right\}$$

6. Apply the following formulas to the data presented in the text to compute the net probability of type 2 for CVR deaths in [65, 70):

$$_nq_{x\cdot\alpha} = (1 - \tfrac{1}{2}{}_nd_{x,\alpha}/{}_nd_{x+})^{-1}({}_nq_x - {}_nd_{x,\alpha}/{}_nd_{x+})$$

$$1 - {}_nq_{x\cdot\alpha} = (1 - {}_nq_x)^{1 - ({}_nd_{x,\alpha}/{}_nd_{x+})}$$

and

$$_nq_{x\cdot\alpha} = 1 - \exp[-n({}_nM_x - {}_nM_{x,\alpha})]$$

7. Keyfitz (1977) has shown that the function

$$H_\alpha = -\int_0^\infty l(x)\ln[l_\alpha(x)]\,dx \bigg/ \int_0^\infty l(x)\,dx$$

where $l_\alpha(x)$ is the survival function when C_α is the sole cause of death, is useful in expressing the change in life expectancy by partial or total elimination of causes of death C_α. If the force of mortality due to C_α changes from $\mu_\alpha(x)$ to $(1 + \delta)\mu_\alpha(x)$, other things remaining the same, the expectation of life at birth in the population changes to $[1 - \delta H_\alpha]$ times the original. A computational formula for H_α is

$$H_\alpha = 1 - \frac{1}{\mathring{e}_0}\sum \frac{l_{x,\alpha}l_x}{M_{x,\alpha} + M_{x+}}\{1 - \exp[-(M_{x,\alpha} + M_{x+})]\}$$

where the summation is for the entire age range from age 0 upwards until the end of the lifespan. Analogously one may define $H_\alpha(a)$ by replacing the life expectancy at birth by life expectancy at age a and restricting the summation for the age range from age a to ∞. Apply this latter formula for CVR using the data presented in the text. [ANSWER: See Table 6.5 and the accompanying calculations.]

Bibliographic Notes

Preston et al. (1972) and Alderson (1981) are useful references for discussions concerning problems encountered in classifying causes of death.

Preston et al. (1972) have compiled an extensive set of multiple-decrement life tables covering many countries over long periods of time. Elandt–Johnson and Johnson (1980) have presented multiple-decrement life tables based on the mortality experience of U. S. white males in 1970. Manton and Stallard (1984) have devoted about a third of their book to cause-of-death analysis. They use concepts from the stochastic compartment model to develop life table methods for the analysis of cause-specific mortality. They have estimated the life expectancy gains attributable to cause elimination, assuming an independent competing-risk setup. They have also examined the impact of prevention of the onset of disease as well as that of a delay in the time of death, given that a person has a chronic disease. Disease dependency has been examined using multiple-cause data.

Chapter 7 | Multiple-Decrement Life Tables II: Analysis of Follow-Up Data

We now shift attention to follow-up data. Imagine that a certain number of individuals are followed over time until each provides a pair of observations (T, C), where T is failure or censoring time and C is cause of failure, if applicable. For censored cases C is left blank; for all other cases it is a k-category entry. We first consider the ungrouped data setup, to be followed by a discussion of the grouped case.

1 Ungrouped Data: An Example

Consider the data (adapted from Nelson, 1970) shown in Table 7.1 These are from a life-testing experiment on an electric appliance. Each appliance, in a sample of size 36, was operated repeatedly by an automatic testing machine. The failure time stands for the number of cycles of use completed before failure. In the original data, 18 different causes of failure were distinguished, bearing codes 1 through 18. For our present purpose, we have collapsed this polytomy into a dichotomy, designating the original code 9 by new code 1, and the remaining original codes by the collapsed (new) code 2.

The entries in the column 6 constitute our estimate of the survival function, where all causes of failure are combined. To estimate the cause-specific survival functions, we use the formula

$$s_\alpha(t) = \sum_{i:t_i \geq t} \left(\frac{D_{i\alpha}}{N_i}\right) s(t_i) \tag{7.1}$$

where s_α stands for an estimate of survival function S_α for cause C_α, s stands for an estimate of the survival function for all causes of failure

108

Table 7.1

Analysis of Failure Times from a Life-Testing Experiment on an Electric Appliance

Cycle of failure	Cause of failure	Number at risk	Number of failures		Estimated survival function		
			Cause 1	Cause 2	Both causes combined	Cause 1	Cause 2
11	2	36	0	1	$\frac{36}{36}$	0.509	0.451
35	2	35	0	1	$\frac{35}{36}$	0.509	0.424
49	2	34	0	1	$\frac{34}{36}$	0.509	0.396
170	2	33	0	1	$\frac{33}{36}$	0.509	0.368
329	2	32	0	1	$\frac{32}{36}$	0.509	0.340
381	2	31	0	1	$\frac{31}{36}$	0.509	0.312
708	2	30	0	1	$\frac{30}{36}$	0.509	0.285
958	2	29	0	1	$\frac{29}{36}$	0.509	0.257
1062	2	28	0	1	$\frac{28}{36}$	0.509	0.229
1167	1	27	1	0	$\frac{27}{36}$	0.509	0.201
1594	2	26	0	1	$\frac{26}{36}$	0.481	0.201
1925	1	25	1	0	$\frac{25}{36}$	0.481	0.174
1990	1	24	1	0	$\frac{24}{36}$	0.454	0.174
2223	1	23	1	0	$\frac{23}{36}$	0.426	0.174
2327	2	22	0	1	$\frac{22}{36}$	0.398	0.174
2400	1	21	1	0	$\frac{21}{36}$	0.398	0.146
2451	2	20	0	1	$\frac{20}{36}$	0.370	0.146
2471	1	19	1	0	$\frac{19}{36}$	0.370	0.118
2551	1	18	1	0	$\frac{18}{36}$	0.343	0.118
2565	—	17	0	0	$\frac{17}{36}$	0.315	0.118
2568	1	16	1	0	$\frac{17}{36}$	0.315	0.118
2694	1	15	1	0	$\frac{15}{16} \times \frac{17}{36}$	0.285	0.118
2702	2	14	0	1	$\frac{14}{16} \times \frac{17}{36}$	0.256	0.118
2761	2	13	0	1	$\frac{13}{16} \times \frac{17}{36}$	0.256	0.089
2831	2	12	0	1	$\frac{12}{16} \times \frac{17}{36}$	0.256	0.059
2034	1	11	1	0	$\frac{11}{16} \times \frac{17}{36}$	0.256	0.030
2059	2	10	0	1	$\frac{10}{16} \times \frac{17}{36}$	0.226	0.030
2112	1	9	1	0	$\frac{9}{16} \times \frac{17}{36}$	0.226	0.0
2214	1	8	1	0	$\frac{8}{16} \times \frac{17}{36}$	0.198	0.0
2478	1	7	1	0	$\frac{7}{16} \times \frac{17}{36}$	0.167	0.0
2504	1	6	1	0	$\frac{6}{16} \times \frac{17}{36}$	0.138	0.0
2829	1	5	1	0	$\frac{5}{16} \times \frac{17}{36}$	0.108	0.0
6367	—	4	0	0	$\frac{4}{16} \times \frac{17}{36}$	0.079	0.0
6976	1	3	1	0	$\frac{4}{16} \times \frac{17}{36}$	0.079	0.0
7846	1	2	1	0	$\frac{2}{3} \times \frac{4}{16} \times \frac{17}{36}$	0.039	0.0
13403	—	1	0	0	$\frac{1}{3} \times \frac{4}{16} \times \frac{17}{36}$	0.0	0.0

Adapted from Nelson (1970).

combined, $D_{i\alpha}$ is the number of failures at time t_i due to cause C_α, and N_i is the number of items exposed to the risk of failure just before the failure time t_i. The marginal distribution of C has probabilities

$$\pi_\alpha = \Pr(C = \alpha) = S_\alpha(0) \tag{7.2}$$

In the present case our estimates of these parameters are $\hat{\pi}_1 = 0.509$, and $\hat{\pi}_2 = 0.451$. Notice that these do not sum to unity. We may therefore prefer the following adjusted estimates:

$$\hat{\pi}_1^* = (0.509)/(0.509 + 0.451) = 0.530; \qquad \hat{\pi}_2^* = (0.451)/(0.509 + 0.451)$$

Given these estimates, we can estimate conditional probabilities such as $\Pr(C = 1 \mid T \geq t)$. Thus from Table 7.1, we have

$$\Pr(C = 1 \mid T \geq 1900) = s_1(1900)/s(1900) = (0.482)/(0.694) = 0.695$$

2   Grouped Data

Imagine that the time axis is partitioned into $(m + 1)$ intervals as

$$[a_0, a_1), \quad [a_1, a_2), \quad \ldots, \quad [a_m, a_{m+1})$$

with $a_0 = 0$ and $a_{m+1} = \infty$. In this setup, a_m represents the upper limit of observation. For each member of a random sample of individuals, we know in which interval the failure time or censoring time falls, and corresponding to each failure time we also know the associated cause of failure, indicated by one or another category of a polytomy, with, say, k categories. For convenience of exposition, we assume that

$$a_i - a_{i-1} = n, \qquad i = 1, \ldots, m \tag{7.3}$$

Let us define the quantities N_i, the number of individuals at risk at time a_i, $D_{i\alpha}$, the number of individuals who failed due to cause C_α in $[a_i, a_{i+1})$, and W_i, the number of individuals censored during the interval $[a_i, a_{i+1})$. Clearly,

$$N_{i+1} = N_i - (D_{i1} + \cdots + D_{ik}) - W_i \tag{7.4}$$

and

$$N_0 = \text{the sample size} \tag{7.5}$$

Writing

$$N_i^* = N_i - \tfrac{1}{2} W_i \tag{7.6}$$

and

$$D_{i+} = D_{i1} + \cdots + D_{ik} \tag{7.7}$$

the standard life table estimate of the probability of failure in $[a_i, a_{i+1})$, when all causes of failure are combined into one category, is

$$\hat{q}_i = D_{i+}/N_i^* \qquad (7.8)$$

This formula applies only if $N_i > 0$. When $N_i = 0$, $\hat{q}_i$ is taken to be unity for convenience.

Obviously, the $\hat{q}_i$ are subject to sampling variation. The following formula suggested by Greenwood (1926) is widely used for estimating the sampling variance of $\hat{q}_i$:

$$\text{var}(\hat{q}_i) = \hat{q}_i(1 - \hat{q}_i)/N_i^* \qquad (7.9)$$

where $\text{var}(\hat{q}_i)$ stands for "an estimate of the variance of $\hat{q}_i$."

An estimate of the crude probability that an individual will fail in $[a_i, a_{i+})$ due to cause C_α, in the presence of all causes of failure, is

$$D_{i\alpha}/N_i^* \qquad (7.10)$$

Once these estimates are on hand, a life table can be constructed in the usual fashion, starting with an arbitrary radix.

Example. Consider the data [adapted from Potter (1967)] shown in Table 7.2. These are from a follow-up study of married women fitted with an intra-uterine device (IUD). The length of the study period was 12 months. A woman could enter the study any time during this period. Failure time in

Table 7.2

Data from a Follow-Up Study of Women Fitted with Intra-Uterine Devices

Interval in months $[a_i, a_{i+1})$	Number at risk	Number of terminations			Number of withdrawals
		Pregnancy	Expulsion	Removal	
[0, 1)	6645	11	167	228	124
[1, 2)	6115	26	106	102	145
[2, 3)	5736	25	95	110	69
[3, 4)	5437	30	82	110	67
[4, 5)	5148	34	56	77	108
[5, 6)	4873	18	44	64	114
[6, 7)	4633	29	43	81	214
[7, 8)	4266	23	42	66	141
[8, 9)	3994	15	38	56	127
[9, 10)	3758	22	28	52	169
[10, 11)	3487	15	23	53	217
[11, 12)	3179	12	16	51	292

this case represents the interval between the time of insertion of an IUD and the time of termination of its use, due to (a) pregnancy, (b) expulsion, or (c) removal for medical or other reasons. If the woman was lost to follow-up or she completed the study period with the IUD in place and without getting pregnant, she was treated as a withdrawn case.

The sample size was 6645 (N_0). As shown in Table 7.2, 406 women "lost the device" during the first month of observation (11 due to pregnancy, 167 because of expulsion, and 228 because of removal), and 124 were withdrawn (i.e., were lost to follow-up or reached the end of the study period with the device in place.) This meant there were

$$6115 = 6645 - 11 - 167 - 228 - 124$$

women with failure times greater than or equal to one month. Similarly,

$$5736 = 6115 - 26 - 106 - 102 - 145$$

had failure times greater than or equal to two months, and so on.

In Table 7.3 the estimated crude probabilities of failure associated with each cause are shown. Notice that

$$N_0^* = N_0 - \tfrac{1}{2}W_0 = 6645 - \tfrac{1}{2}124 = 6583.0$$
$$N_1^* = N_1 - \tfrac{1}{2}W_1 = 6115 - \tfrac{1}{2}145 = 6042.5$$

Table 7.3

Crude Probabilities of Failure from Specific Causes

Interval in months	Effective number at risk	Crude probabilities of failure in the interval			
		Pregnancy	Expulsion	Removal	All causes
[0, 1)	6583.0	0.001671	0.025368	0.034635	0.061674
[1, 2)	6042.5	0.004303	0.017542	0.016880	0.038726
[2, 3)	5701.5	0.004385	0.016662	0.019293	0.040340
[3, 4)	5403.5	0.005552	0.015175	0.020357	0.041084
[4, 5)	5094.5	0.006675	0.010993	0.015116	0.032784
[5, 6)	4816.0	0.003738	0.009136	0.013289	0.026163
[6, 7)	4526.0	0.006407	0.009501	0.017897	0.033805
[7, 8)	4195.5	0.005482	0.010011	0.015731	0.031224
[8, 9)	3930.5	0.003816	0.009668	0.014248	0.022732
[9, 10)	3678.5	0.005989	0.007622	0.014155	0.027766
[10, 11)	3378.5	0.004440	0.006808	0.015657	0.026535
[11, 12)	3033.0	0.003956	0.005275	0.016815	0.026947

and so on. With these numbers (commonly called the effective numbers at risk) on hand, the crude probability of failure due to pregnancy in the first interval was computed as $\frac{11}{6583} = 0.001671$, and the corresponding numbers for the other two causes were obtained as $\frac{167}{6583}$ and $\frac{228}{6583}$. The last column of the table shows the sum of the estimated crude probabilities for the various causes of failure. Thus

$$0.061674 = 0.001671 + 0.025368 + 0.034635$$

These are the values of $\hat{q}_i$ [see Eq. (7.8)]. A multiple-decrement life table constructed on the basis of the crude probabilities of failure (Table 7.3) is presented in Table 7.4 with a radix of 100,000. In column 2 of Table 7.4 are shown the numbers of women in the hypothetical cohort (initially of size 100,000) who are expected to complete one month, two months, etc. with the IUD in place. Thus out of the original 100,000 women, 93,833 are expected to retain the device at least one month, 90,199 are expected to do so at least two months, and so on. [Notice that the figures in this column satisfy the recurrence relation: the number in the ith row equals the number above it minus d_i to the latter's right. Thus, $93,833 = 100,000 - 6167$; $90,199 = 93833 - 3634$; and so on.] Column 3 was completed using the formula

Table 7.4

Multiple-Decrement Life Table Constructed on the Basis of the Figures Presented in Table 7.3: Expected Number of Deaths from Each Cause and All Causes Combined in Each Duration

Interval in months	Expected number of survivors to the beginning of interval	Expected number of deaths within interval			
		All causes	Pregnancy	Expulsion	Removal
[0, 1)	100000	6167	167	2537	3463
[1, 2)	93833	3634	404	1646	1584
[2, 3)	90199	3639	396	1503	1740
[3, 4)	86560	3555	481	1312	1762
[4, 5)	83005	2721	554	912	1255
[5, 6)	80284	2100	300	733	1067
[6, 7)	78184	2643	501	743	1399
[7, 8)	75541	2358	414	756	1188
[8, 9)	73183	2030	279	708	1043
[9, 10)	71153	1975	426	542	1007
[10, 11)	69178	1863	307	471	1085
[11, 12)	67315	1753	266	355	1132
[12, ∞)	65562				

$d_i = l_i \hat{q}_i$, more commonly expressed as $d_x = l_x q_x$. Thus

$$d_0 = 100{,}000 \times 0.061674 = 6167; \qquad d_1 = 93833 \times 0.038726 = 3634$$

The numbers in the remaining columns were computed by applying the estimated cause-specific crude probabilities of failure to the numbers in column 2. Thus, under pregnancy we have $167 = 100{,}000 \times 0.001671$; under expulsion we have $1646 = 93833 \times 0.017542$; and so on.

3　Cumulative Probability of Failure

In the multiple-decrement life table for follow-up data, a quantity of special interest is the cumulative probability of failure over time. Thus, with reference to the IUD data, we may be interested in the probability that a woman will lose the device due to expulsion within t months of insertion. Table 7.5 contains estimates of these probabilities for each cause and for all causes combined. The procedure involved in the computation of these figures is simple: we simply cumulate the number in the d_{i+} and $d_{i,\alpha}$ columns of Table 7.4 and divide the results by the radix. The last row of Table 7.5 shows that 34.4 percent of women fitted with an IUD may be expected to lose the device within 12 months of insertion (4.5 percent due to pregnancy, 12.2 percent due to expulsion, and 17.7 percent due to removal.)

Table 7.5

Probability of Failure within the Interval [0, a) Estimated from the
Figures in Table 7.4

Interval in months	Probability of losing the device within [0, a)			
	All causes	Pregnancy	Expulsion	Removal
[0, 1)	0.06167	0.00167	0.02537	0.03463
[1, 2)	0.09801	0.00571	0.04183	0.05047
[2, 3)	0.13440	0.00967	0.05686	0.06787
[3, 4)	0.16995	0.01446	0.06998	0.08549
[4, 5)	0.19716	0.02002	0.07910	0.09804
[5, 6)	0.21816	0.02302	0.08643	0.10871
[6, 7)	0.24459	0.02803	0.09386	0.12270
[7, 8)	0.26817	0.03217	0.10142	0.13458
[8, 9)	0.28847	0.03496	0.10850	0.14501
[9, 10)	0.30822	0.03982	0.11392	0.15508
[10, 11)	0.32685	0.04229	0.11863	0.16593
[11, 12)	0.34438	0.04495	0.12218	0.17725

Obviously, we can similarly compute the probability of failure within the interval $[a, b)$, for $b > a$, given survival until a. We aggregate the d_{i+} and $d_{i,\alpha}$ numbers over the interval $[a, b)$ and divide the resulting figures by l_a. Thus from Table 7.4, we have 3575 for the sum of $d_{i,\text{expulsion}}$, over the interval $[6,12)$, and $l_6 = 78184$. Therefore $\frac{3575}{78184} = 0.0457$ is our estimate of the probability that a woman will lose the device due to expulsion during the last six months of the first year, given that she has retained it until the end of the first six months. This figure may be compared to 0.08643 for the first six months.

Standard Errors

To estimate the standard errors of the statistics mentioned above, the following formulas may be used:

a. Writing $\text{var}(\hat{q}_i)$ for an estimate of the variance of $\hat{q}_i$, we use the binomial formula

$$\text{var}(\hat{q}_i) = \hat{q}_i(1 - \hat{q}_i)/N_i^*$$

Thus from Tables 7.3 and 7.5, for the estimated probability of losing the device during the first month after insertion, we have

$$\text{var}(\hat{q}_0) = 0.06167 \times 0.938326/6583.0 = 8.7903 \times 10^{-6}$$

and taking the square root the corresponding standard error is obtained as 2.965×10^{-3}. Notice that the standard errors of $\hat{p}_i$ are the same as those of $\hat{q}_i$. In column 2 of Table 7.6, the standard errors thus estimated for various values of i are shown.

b. Similarly, we estimate the variance of $\hat{q}(i, \alpha)$, the estimated probability of losing the device due to cause C_α, using the formula

$$\text{var}[\hat{q}(i, \alpha)] = \hat{q}(i, \alpha)[1 - \hat{q}(i, \alpha)]/N_i^*$$

Thus from Tables 7.3 and 7.5

$$\text{var}[\hat{q}(0, \text{expulsion})] = 0.025368 \times 0.974632/6583.0 = 3.7558 \times 10^{-6}$$

giving a standard error of 1.938×10^{-3}. The numbers thus computed for various values of i are shown in column 3 of Table 7.6.

c. For the cumulative proportion of women losing the device within the time interval $[0, a)$, we apply the following formula:

$$\text{var}(\hat{P}_{0a}) = \hat{P}_{0a}^2 \sum_0^{a-1} \frac{\hat{q}_h}{(1 - \hat{q}_h)N_h^*}$$

Thus from Tables 7.3 and 7.5, for $a = 2$, we have for all causes combined

Table 7.6

Estimated Standard Errors

Interval	$\hat{q}_i{}^a$	$\hat{q}(i, \text{expulsion})^b$	$\hat{P}(0, a)^{c,d}$	$\hat{Q}(0, a, \text{expulsion})^e$
(1)	(2)	(3)	(4)	(5)
[0, 1)	0.0030	0.0019	0.0030	0.0019
[1, 2)	0.0025	0.0017	0.0037	0.0025
[2, 3)	0.0026	0.0017	0.0042	0.0029
[3, 4)	0.0027	0.0017	0.0047	0.0032
[4, 5)	0.0025	0.0015	0.0050	0.0034
[5, 6)	0.0023	0.0014	0.0052	0.0035
[6, 7)	0.0027	0.0014	0.0054	0.0037
[7, 8)	0.0027	0.0015	0.0056	0.0038
[8, 9)	0.0026	0.0016	0.0058	0.0040
[9, 10)	0.0027	0.0014	0.0060	0.0041
[10, 11)	0.0028	0.0014	0.0061	0.0042
[11, 12)	0.0029	0.0013	0.0063	0.0043

[a] q_i stands for the probability of losing the device in the ith interval, all causes combined.

[b] $q(i, \text{expulsion})$ stands for the crude probability of losing the device in the ith interval, due to expulsion, when all causes are in operation.

[c] $P(0, a)$ stands for the cumulative probability of retaining the device up to the beginning of the interval $[a, a + 1)$.

[d] Note that variance of $P(0, a)$ is the same as the variance of $[1 - P(0, a)]$.

[e] $Q(0, a, \text{expulsion})$ stands for the cumulative probability of losing the device due to expulsion in the interval $[0, a)$.

$$\text{var}(\hat{P}_{02}) = 1 - (0.09801)^2[0.061674/(0.938326 \times 6583)$$
$$+ 0.038726/(0.961274 \times 6042.5)]$$
$$= 0.813586 \times 1.665 \times 10^{-5}$$

giving the corresponding standard error as 3.68×10^{-3}. The numbers thus computed for various values of a are shown in column 4 of Table 7.6.

d. Finally, to estimate the variance of the estimated cumulative proportion, say $\hat{Q}(0, a, \alpha)$, of losing the device within $[0, a)$ due to a specific cause, say, C_α, we use the formula

$$\text{var}[\hat{Q}(0, a, \alpha)] = \sum_0^{a-1} A_i^2 \, \text{var}(\hat{p}_i) + \sum_0^a B_i^2 \, \text{var}[\hat{q}(i, \alpha)]$$
$$+ 2 \sum_0^{a-1} A_i B_i \, \text{cov}[\hat{q}(i, \alpha), \hat{p}_i]$$

where

$$A_i = \frac{1}{p_i} \sum_{i+1}^{a} (\hat{p}_0 \cdots \hat{p}_{j-1})\hat{q}(j, \alpha)$$

$$B_i = \hat{p}_0 \cdots \hat{p}_{i-1}, \qquad \text{with } B_0 = 1$$

and

$$\text{cov}[\hat{q}(i, \alpha), \hat{p}_i] = -\hat{q}(i, \alpha)\hat{p}_i/N_i^*$$

The standard errors (square root of variance) estimated using this formula with specific reference to losing the device because of expulsion are shown in the last column of Table 7.6.

Problems and Complements

1. Consider the policy holders of a life insurance company. Their number may be decremented by death, disability, or lapse (because of failure to pay a due premium), and incremented by new policies issued. Suppose initially there were 200 policy holders and that the following events are reported to have occurred in sequence: one death, one lapse, one increment, one death, one death, one disability, one increment, one disability, one lapse, one increment, one death, one death. Complete the following table and then estimate the cause-specific survival distributions assuming independence of the causes of decrements.

Event	Number before the event	Number after the event
Death	200	199
Lapse		
Increment		
Death		
Death		
Disability		
Increment		
Disability		
Lapse		
Increment		
Death		
Death		

2. Do you think the assumption that the causes of decrements are independent is tenable in the context described in Exercise 1? For instance, do you think the probability that an individual will let his (her) policy lapse is the same irrespective of whether he (she) is on the brink of death or disability?

3. Perform a competing risk analysis of the following (hypothetical) data from a follow-up study:

Years of follow-up $[t_i, t_{i+1})$	Number living at t_i	Number of deaths in $[t_i, t_{i+1})$		Withdrawals
		Cause 1	Cause 2	
[0, 1)	600	120	20	60
[1, 2)	400	60	10	50
[2, 3)	280	20	5	45
[3, 4)	210	10	4	36
[4, 5)	160	6	3	31
[5, 6)	120	3	4	23
[6, 7)	90	2	2	16
[7, 8)	70	1	1	18
[8, 9)	50	1	1	13
[9, 10)	35	1	1	8
[11, 12)	25	1	0	9
[12, 13)	15	0	0	15

Table 7.7

Number of Separations, Widowhoods, and Intact Marriages at Given Durations since First Marriage: U. S. White Women Who First Got Married between January 1930 and December 1959

Duration	Separations	Widowhoods	Intact marriages
[0, 1)	287	43	465
[1, 2)	231	33	578
[2, 3)	211	37	632
[3, 4)	200	44	700
[4, 5)	156	54	736
[5, 6)	126	33	646
[6, 7)	112	33	673
[7, 8)	115	28	669
[8, 9)	102	36	629
[9, 10)	68	33	758
[10, 11)	60	18	649
[11, 12)	55	28	749
[12, 13)	54	22	724
[13, 14)	42	30	867
[14, 15)	48	16	616
[15, 16)	28	28	468
[16, 17)	30	31	517
[17, 18)	19	24	617
[18, 19)	15	27	586
[19, 20)	14	26	567
[20, ∞)	46	114	3656

4. The data in Table 7.7 pertain to marital status, as of January 1960, of U. S. white women who first got married between January 1930 and December 1959. Only two ways of exit from the married state are recognized: (i) separation and (ii) widowhood. Intact marriages as of January 1960 are treated as censored cases. Estimate for $i = 0, 1, \ldots, 19$:

    a. the probability of marital dissolution in duration $[i, i + 1)$;
    b. the probability of separation in duration $[i, i + 1)$;
    c. the probability of widowhood in duration $[i, i + 1)$;
    d. the (cumulative) probability of marital dissolution by duration $i + 1$;
    e. the standard error of the estimate in (d);
    f. the (cumulative) probability of separation by duration $i + 1$;
    g. the standard error of the estimate in (f);
    h. the (cumulative) probability of widowhood by duration $i + 1$;
    i. the standard error of the estimate in (h);
    j. the probability of marital dissolution in duration $[i, i + 1)$, if widowhood never occurs (i.e., if exit via widowhood is eliminated);
    k. the (cumulative) probability of marital dissolution by duration $i + 1$, if widowhood never occurs;
    l. the standard error of the estimate in (k);
    m. the median length of marriage for those who exit via separation within 20 years of marriage.

5. The data in Table 7.8 are from a sample survey conducted in 1973 and pertain to U. S. white women who got married a second time before age 40 and who had had at least one child in the first marriage. For each duration since second marriage, Table 7.8 shows the number of women who had had at least one live birth in the second marriage, and, among those who did not have a live birth in the second marriage (as of the interview date), the number separated or widowed, and the number of intact (second) marriages. Using the data estimate

    a. the (cumulative) probability of a live birth in the second marriage by given duration in months, since second marriage;
    b. the standard error of the estimate in (a);
    c. the (cumulative) probability of marital dissolution (via separation or widowhood) before a live birth occurred in the second marriage, by duration in months, since second marriage.
    d. the standard error of the estimate in (c).

Bibliographic Notes

Follow-up data subject to two or more causes of failure are very common in life-testing work. In the literature, a distinction is usually made between what are referred to as "series systems" and "parallel systems." The former refers to entities (e.g., mechanical devices) that fail if at least one part fails, whereas a parallel system refers to an entity that fails if and only if all parts fail (Elandt–Johnson and Johnson, 1980).

The competing risk model is a special case of random censoring. [Random censoring usually occurs when a life test is performed on biological subjects. Here the volume and the times of censoring are not under the control of the investigator—censoring occurs at random points in time.] The model for random censorship was proposed by Effron (1967). Methods for the analysis of data from randomly censored life tests are expounded in Breslow and Crowley (1974). Reviews of the literature on competing risks are available in Birnbaum (1979), David and Moeschberger (1978), and Elandt–Johnson and Johnson (1980).

Table 7.8

Number of Second Marriages of Given Duration with at Least One Live Birth
(in the Second Marriage) and Those of Given Duration with No Live Birth in
the Second Marriage Classified into Intact and Dissolved Marriages (the Latter
via Separation or Divorce): U. S. Whites, 1973

Second marriage duration (months)	Number with at least one live birth in second marriage	Number with no live birth in second marriage	
		Marriage dissolved	Marriage intact
[6, 12)	57	1	16
[12, 18)	55	5	17
[18, 24)	34	2	10
[24, 36)	31	2	31
[36, 48)	15	3	16
[48, 60)	3	4	5
[60, 72)	4	1	9
[72, 84)	1	0	15
[84, 96)	1	0	4
[96, ∞)	0	1	24

Answer:

	Duration (months)								
	12	18	24	36	48	60	72	84	96
(a)	0.159	0.321	0.427	0.538	0.604	0.619	0.642	0.649	0.659
(b)	0.019	0.025	0.025	0.026	0.026	0.028	0.036	0.039	0.041
(c)	0.003	0.018	0.023	0.031	0.044	0.064	0.070	0.070	0.070
(d)	0.003	0.007	0.008	0.010	0.012	0.015	0.015	0.015	0.015

Bibliographic Notes (*cont.*). In the 1960s, Potter (1969) developed a life table method to analyze the use-effectiveness of intra-uterine devices as a competing risk problem. Later Tietze and Lewit (1973) recommended somewhat different procedures for the statistical evaluation of intra-uterine contraception. The Potter method, applied to a large data base assembled in the United States, produced estimates of termination rates different from that produced by the application of the other method (Chandrasekharan and Hermalin, 1975). Jain and Sivin (1977) investigated the reasons for this discrepancy and found that a substantial fraction of acceptors had incomplete observations, and neither of the two methods seemed to be able to effectively deal with the resulting setup.

Chapter 8 | Multiple-Decrement Life Tables III: General Theory

In this chapter we briefly discuss the probability theory underlying multiple-decrement life tables. For more details the reader is referred to Chiang (1968), Altschuler (1970), Prentice *et al.* (1978), David and Moeschberger (1978), Kalbfleisch and Prentice (1978), Birnbaum (1979), and Elandt-Johnson and Johnson (1980).

The exposition is organized around causes of death. It could as well be organized around cause-specific failure types. When discussing cause of death, it should be recognized that official death certificates in countries such as the United States contain information on what are often referred to as the immediate, intervening, underlying, and contributory causes of death. But more often than not only the underlying cause of death is analyzed. [The notion that death may be the outcome of the joint action of two or more causes remains to be satisfactorily operationalized in human mortality analysis.] No attempt is made herein to consider more than the underlying cause of death.

1 Some Basic Notions

Suppose that in a population there are k possible causes of death. It is mathematically convenient to attach to each cause of death a random variable representing what may be called the potential (latent) age at death due to that cause. Let $X_1, \ldots, X_k$ be these random variables. Then we define their joint survival function as

$$S(x_1, \ldots, x_k) = \Pr[(X_1 > x_1) \cap \cdots \cap (X_k > x_k)] \qquad (8.1)$$

The marginal survival functions are defined as

$$S_\alpha(x_\alpha) = S(0, \ldots, 0, x_\alpha, 0, \ldots, 0) \tag{8.2}$$

As already mentioned, we assume that each death is caused by a single cause. This implies that we cannot observe $X_1, \ldots, X_k$ jointly. In fact we observe only one of them, whichever is realized first. That is we observe only

$$X = \min(X_1, \ldots, X_k) \tag{8.3}$$

We call

$$S(x, \ldots, x) = \Pr(X > x) = \Pr[(X_1 > x) \cap \cdots \cap (X_k > x)] \tag{8.4}$$

the *overall* survival function. This represents the probability that death will not occur before age x, given that all causes are in operation. To fix ideas let us consider the following example of a joint survival function (Tsiatis, 1975):

$$S(x_1, x_2) = \exp(-\beta_1 x_1 - \beta_2 x_2 - \gamma x_1 x_2) \tag{8.5}$$

where $\beta_1 > 0$, $\beta_2 > 0$, and $0 \le \gamma \le \beta_1 \beta_2$. The two marginal survival functions are in this case

$$S_1(x_1) = \exp(-\beta_1 x_1) \tag{8.6}$$

and

$$S_2(x_2) = \exp(-\beta_2 x_2) \tag{8.7}$$

The overall survival function is

$$S(x) = S(x, x) = \exp(-\beta_1 x - \beta_2 x - \gamma x^2) \tag{8.8}$$

Notice that $S(x, x)$ is a function of a single variable x. We may therefore refer to it simply as $S(x)$. We now define two types of force of mortality, one type corresponding to $S(x_1, \ldots, x_k)$ and the other corresponding to $S(x)$. There are k forces of mortality of the former type, one for each cause of death. For the survival function given in (8.5), we have [taking partial derivatives of the negative of the natural logarithm of $S(x_1, x_2)$, and then evaluating it at $x_1 = x_2 = x$],

$$\mu(x, 1) = \beta_1 + \gamma x \tag{8.9}$$

and

$$\mu(x, 2) = \beta_2 + \gamma x \tag{8.10}$$

These represent the instantaneous rates of dying from the respective causes at age x, given survival until age x, assuming that the survival function is as specified in (8.5), and that both causes are in operation. It is easily verified that the force of mortality corresponding to the overall survival function (8.8) equals the sum of the two cause-specific *crude* forces of mortality (8.9) and (8.10). This additive property holds in general. That is, the overall force of mortality is equal to the sum of what may be regarded as the *component crude forces*:

$$\mu(x) = \mu(x, 1) + \mu(x, 2) \tag{8.11}$$

The so-called *net* force of mortality for cause C_α corresponds to the marginal survival function for that cause. Using the notation $\lambda(x, \alpha)$ for this type of force of mortality for cause C_α, we have for the particular case given above [see (8.6) and (8.7)]

$$\lambda(x, 1) = \beta_1 \tag{8.12}$$

and

$$\lambda(x, 2) = \beta_2 \tag{8.13}$$

This example illustrates the point that the crude force of mortality is not necessarily equal to the net force of mortality. [In this example, the two will be equal, for each cause, if and only if $\gamma = 0$.]

2   Expected Proportion of Deaths from a Given Cause

One practical question of interest concerns the probability that a newborn individual survives to age x and then dies from a given cause, say C_α. The conditional probability of death from C_α in the age interval $(x, x + \Delta x)$, given survival until age x, when all causes of death are simultaneously in operation, is approximately $\mu(x, \alpha) \Delta x$. The unconditional probability of dying from cause C_α in $(x, x + \Delta x)$ is therefore $S(x)\mu(x, \alpha) \Delta x$. Now let us define an indicator variable J such that it takes the value α if the cause of death is C_α and zero otherwise. The crude probability distribution of age at death from cause C_α is given by

$$\Pr[(X \leq x) \cap (J = \alpha)] = \int_0^x S(a)\mu(a, \alpha) \, da \tag{8.14}$$

Obviously,

$$\Pr[X \leq x] = \Pr[(X \leq x) \cap (J = 1)] + \cdots + \Pr[(X \leq x) \cap (J = k)] \tag{8.15}$$

Now, if we write π_α for the limit of (8.14) as x tends to ∞, then from (8.15) we have

$$\pi_1 + \cdots + \pi_k = 1 \tag{8.16}$$

Clearly,

$$\pi_\alpha = \Pr[J = \alpha] \tag{8.17}$$

is the expected proportion of deaths due to cause C_α.

3   The Distribution of Age at Death

If we denote $\Pr[(X > x) \cap (J = \alpha)]$ by the symbol $P^*(x, \alpha)$ and use the symbol $Q^*(x, \alpha)$ for $\Pr[(X < x) \cap (J = \alpha)]$, then we notice that

$$\pi_\alpha = Q^*(\infty, \alpha) = P^*(0, \alpha) \tag{8.18}$$

where for convenience we write $Q^*(\infty, \alpha)$ for the limit of $Q^*(x, \alpha)$ as x tends to ∞ and $P^*(0, \alpha)$ for the limit of $P^*(x, \alpha)$ as x tends to 0. Furthermore,

$$\Pr(X < x) = Q^*(x, 1) + \cdots + Q^*(x, k) \tag{8.19}$$

and

$$\Pr(X > x) = P^*(x, 1) + \cdots + P^*(x, k) \tag{8.20}$$

There is no guarantee that $P^*(x, \alpha)$ will behave like the usual survival function. $P^*(0, \alpha)$ may not be unity, for example. But we could construct a survival function from each $P^*(x, \alpha)$ by dividing it by $\pi_\alpha = P^*(0, \alpha)$. In other words, we may define

$$S^*(x, \alpha) = (1/\pi_\alpha)P^*(x, \alpha) \tag{8.21}$$

to represent the probability that age at death is greater than x, given that the cause of death is C_α. This incidentally is the survival function among those who die of cause C_α, in the presence of all causes of death. We can estimate it from the usual type of data. The corresponding probability density function is obtained by applying the formula

$$f^*(x, \alpha) = -d[\ln S^*(x, \alpha)]/dx \tag{8.22}$$

which gives

$$f^*(x, \alpha) = (1/\pi_\alpha)\mu(x, \alpha)S(x) = -(1/\pi_\alpha)(dP^*(x, \alpha)/dx) \tag{8.23}$$

We get from (8.23) the expression for $\mu(x, \alpha)$

$$\mu(x, \alpha) = -\frac{1}{S(x)}\frac{dP^*(x, \alpha)}{dx} \tag{8.24}$$

Remembering that $S(x) = \Pr(X > x)$, the following alternative expression for

$\mu(x, \alpha)$ can be obtained from (8.24) by substituting for $S(x)$ from (8.20)

$$\mu(x, \alpha) = -\frac{1}{\Sigma_i P^*(x, i)} \frac{dP^*(x, \alpha)}{dx} \tag{8.25}$$

The force of mortality associated with $S^*(x, \alpha)$ defined in (8.21), on the other hand, is

$$\mu^*(x, \alpha) = \frac{f^*(x, \alpha)}{S^*(x, \alpha)} = -\frac{1}{P^*(x, \alpha)} \frac{dP^*(x, \alpha)}{dx} \tag{8.26a}$$

The ratio of (8.24) to (8.26a) gives

$$\mu(x, \alpha)/\mu^*(x, \alpha) = P^*(x, \alpha)/S(x) \tag{8.26b}$$

which represents the conditional probability of dying from cause C_α, conditional on surviving until age x.

Cause-Specific Components of the Probability of Dying in an Age Interval

Let $q(i, \alpha)$ stand for the probability that an individual who has survived to the beginning of the age interval (x_i, x_{i+1}) will die from cause C_α before reaching the end of that age interval, and let $q(i, +)$ stand for the corresponding probability when the cause of death is left open. We can show that $q(i, +) = q(i, 1) + q(i, 2) + \cdots + q(i, k)$. This follows from the assumption that death is attributable to one and only one of the k causes. Writing $\mu(t, +)$ for the sum $\mu(t, 1) + \cdots + \mu(t, k)$, we have in the setup just described

$$q(i, +) = \int_{x_i}^{x_{i+1}} \mu(t, +) \exp\left[-\int_{x_i}^{t} \mu(s, +)\, ds\right] dt$$

$$= \int_{x_i}^{x_{i+1}} \left[\sum_\alpha \mu(t, \alpha)\right] \exp\left[-\int_{x_i}^{t} \mu(s, +)\, ds\right] dt$$

$$= \sum_\alpha q(i, \alpha) \tag{8.27}$$

A Particular Case

Suppose $\mu(x, \alpha)$ is constant in the interval (x_i, x_{i+1}), say, at $m_{i\alpha}$, for all C_α. Let $m_{i+} = \Sigma_\alpha m_{i\alpha}$. Then it is easily verified that

$$q(i, \alpha) = (m_{i\alpha}/m_{i+})q(i, +) \tag{8.28}$$

and

$$q(i, +) = 1 - \exp[-(x_{i+1} - x_i)m_{i+}] \qquad (8.29)$$

Thus, $q(i, \alpha)$ can be obtained as a share of $q(i, +)$, in proportion to the ratio $m_{i\alpha}/m_{i+}$, i.e., (the force of mortality specific to cause C_α)/(the total force of mortality). Usually one approximates $m_{i\alpha}$ by the age-specific death rate due to cause C_α, for (x_i, x_{i+1}). Then the ratio $m_{i\alpha}/m_{i+}$ is the cause-of-death ratio in the age interval involved. Thus, under the assumption of constant force of mortality, the crude probability of death can be obtained by multiplying the total probability of death by the corresponding cause-of-death ratio.

4   The Probability of Eventually Dying from Cause C_α, at an Age $\geq x$

Let $P(\alpha \mid x)$ stand for the probability of dying from cause C_α after surviving to age x. Then

$$P(\alpha \mid x) = \frac{\Pr[(X > x) \cap (J = \alpha)]}{\Pr(X > x)} = \frac{P^*(x, \alpha)}{S(x)} \qquad (8.30)$$

But

$$P^*(x, \alpha) = \Pr[(X > x) \cap (J = \alpha)] = \int_x^\infty S(t)\mu(t, \alpha)\, dt$$

$$= \sum \int_{x_i}^{x_{i+1}} S(t)\mu(t, \alpha)\, dt$$

$$= \sum \int_{x_i}^{x_{i+1}} \left\{ \exp\left[-\int_0^t \mu(s, +)\, ds \right] \right\} \mu(t, \alpha)\, dt \qquad (8.31)$$

where the summation is over the intervals typified by (x_i, x_{i+1}) into which $[x, \infty)$ is partitioned. It is easily seen that the last member of Eq. (8.31) gives on simplification:

$$P^*(x, \alpha) = \sum l(x_i) \int_{x_i}^{x_{i+1}} \left\{ \exp\left[-\int_{x_i}^t \mu(s, +)\, ds \right] \right\} \mu(t, \alpha)\, dt \qquad (8.32)$$

where

$$l(x_i) = \exp\left[-\int_0^{x_i} \mu(s, +)\, ds \right] \qquad (8.33)$$

Recalling the definition of $q(i, \alpha)$, we can express (8.32) succinctly as

$$P^*(x, \alpha) = \sum l(x_i)q(i, \alpha) \qquad (8.34)$$

We thus have from (8.30), and (8.34),

$$p(x \mid \alpha) = \Sigma \, l(x_i) q(i, \alpha) / l(x) \tag{8.35}$$

where the summation is over the intervals into which $[x, \infty)$ has been partitioned: $[x = x_0, x_1), [x_1, x_2), \ldots, [x_\omega, \infty)$. Life tables portraying current mortality give $l(x)$ values, and $q(i, \alpha)$ can be estimated following the method outlined above [see (8.28) and (8.29)].

5 The Effect of a Particular Cause of Death on Mortality Pattern

The question as to how the mortality pattern might change if a particular cause C_α were "eliminated," has been addressed by many scholars over the past two centuries or so, but a fully satisfactory answer to it has not been found as yet. Chiang (1968), David and Moeschberger (1978), Gail (1975), and others have discussed the problem under the broad heading of competing risks. An exhaustive review of the related literature is beyond the scope of the present work. We mention below a few developments.

Net Probability

Let $r_{i\alpha}$ be the probability that a person alive at the beginning of the age interval (x_i, x_{i+1}) will die before reaching the end of the interval, when C_α is the only cause of death in operation. Then by definition

$$r_{i\alpha} = \int_{x_i}^{x_{i+1}} \mu(t, \alpha) \exp\left[-\int_{x_i}^{t} \mu(s, \alpha) \, ds \right] dt \tag{8.36}$$

We mention below a number of ways of estimating these net probabilities.

Constant Force of Mortality

Let us assume that the force of mortality is constant, say, equal to $m_{x\alpha}$ in the age interval $(x, x + n)$. Then

$$r_{i\alpha} = \int_{x}^{x+n} m_{x\alpha} \exp[-(t - x) m_{x\alpha}] \, dt = 1 - \exp(-n m_{x\alpha}) \tag{8.37}$$

In practice, one may approximate $m_{x\alpha}$ by the "observed" age-specific death rate in the age interval $(x, x + n)$ due to cause C_α.

Actuarial Method

Consider a particular cause of death, say C_α, and group all other causes into one residual category, say C_{α^*}. Let us assume that deaths are evenly distributed within each age interval. Further, let us assume that the potential ages at death are statistically independent. Then $q(x, \alpha)$, the probability that an individual who has survived to age x, will die from cause C_α before reaching age $(x + n)$ is given by

$$q(x, \alpha) = \int_0^n \mu(x + t, \alpha) \exp\left[\int_x^{x+t} \mu(s, +)\, ds\right] dt \qquad (8.38)$$

If

$$p(x, x + t) = \exp\left[-\int_x^{x+t} \mu(s, +)\, ds\right] \qquad (8.39)$$

$$p_\alpha(x, x + t) = \exp\left[-\int_x^{x+t} \mu(s, \alpha)\, ds\right] \qquad (8.40)$$

and

$$p_{\alpha*}(x, x + t) = \exp\left[-\int_x^{x+t} \mu(s, \alpha*)\, ds\right] \qquad (8.41)$$

where the last two are the probabilities of surviving from age x until age $(x + t)$, when C_α or $C_{\alpha*}$, as the case may be, is the only cause of death in operation, whereas $p(x, x + t)$ stands for the corresponding probability when both causes are present. Under the setup, we have by virtue of the assumption of independence

$$p(x, x + t) = p_\alpha(x, x + t)p_{\alpha*}(x, x + t) \qquad (8.42)$$

Furthermore, given that deaths are evently distributed, the probability of dying within a segment of an age interval is proportional to the length of that segment. Consequently,

$$p_{\alpha*}(x, x + t) = 1 - q_{\alpha*}(x, x + t) = 1 - (t/n)q_{\alpha*}(x, x + n) \qquad (8.43)$$

where $q_{\alpha*}(x, x + n)$ is the probability of dying in the age interval $(x, x + n)$, given survival until age x, and no cause in operation other than $C_{\alpha*}$. Equation (8.38) now can be expressed as

$$
\begin{aligned}
q(x, \alpha) &= \int_0^n p(x, x + t)\mu(x + t, \alpha)\, dt \\
&= \int_0^n p_\alpha(x, x + t)p_{\alpha*}(x, x + t)\mu(x + t, \alpha)\, dt \\
&= \frac{1}{n}\int_0^n \left[1 - \frac{t}{n}q_{\alpha*}(x, x + n)\right]q_\alpha(x, x + n)\, dt \\
&= q_\alpha(x, x + n)[1 - \tfrac{1}{2}q_{\alpha*}(x, x + n)]
\end{aligned}
\qquad (8.44)
$$

Similarly,

$$q(x, \alpha^*) = q_{\alpha^*}(x, x + n)[1 - \tfrac{1}{2}q_\alpha(x, x + n)] \qquad (8.45)$$

From (8.44)

$$q_\alpha(x, x + n) = \frac{q(x, \alpha)}{1 - \tfrac{1}{2}q_{\alpha^*}(x, x + n)} \qquad (8.46)$$

A simple actuarial estimate which usually performs well is obtained by replacing $q_{\alpha^*}(x, x + n)$ in (8.46) by $q(x, \alpha^*)$, thus giving

$$q_\alpha(x, x + n) = \frac{q(x, \alpha)}{1 - \tfrac{1}{2}q(x, \alpha^*)} \qquad (8.47)$$

A different actuarial estimate is obtained by solving for $q_\alpha(x, x + n)$ from (8.44) and (8.45), after eliminating $q_{\alpha^*}(x, x + n)$. Elimination of $q_{\alpha^*}(x, x + n)$ gives a quadratic equation in $q_\alpha(x, x + n)$ whose solution consistent with $0 \le q(x, \alpha) \le 1$ is

$$q_\alpha(x, x+n)=\tfrac{1}{2}\{2-q(x, \alpha^*)+q(x, \alpha)-\sqrt{[(2-q(x, \alpha^*)+q(x, \alpha)]^2-8q(x, \alpha)}\} \qquad (8.48)$$

Chiang's Estimate

Chiang (1968) proposed an estimate on the basis of the so-called proportionality assumption, namely, that within any age interval each cause-specific force of mortality may vary in absolute magnitude but such that the ratios between those force remain constant. Suppose $\mu(t, \alpha)/\mu(t, +) = k_{i\alpha}$, a constant, for all ages t, in interval x_i to x_{i+1}. Then from (8.37)

$$r_{i\alpha}/r_{i+} = k_{i\alpha} \qquad (8.49)$$

and, consequently,

$$r_{i\alpha} = 1 - \exp\left[-\int_{x_i}^{x_{i+1}} \mu(t, \alpha)\right] dt$$

$$= 1 - \exp\left[-k_{i\alpha}\int_{x_i}^{x_{i+1}} \mu(t, +) \, dt\right]$$

$$= 1 - p_i^{kC_{i\alpha}} \qquad (8.50)$$

where

$$p_i = \exp\left[-\int_{x_i}^{x_{i+1}} \mu(t, +)\right] dt$$

Chiang suggested that as an estimate of $k_{i\alpha}$, the ratio of [see (8.49)] the age-and-cause (α) specific death rate $[M_\alpha(x_i, x_{i+1})]$ to the age-specific (all causes combined) death rate $[M_+(x_i, x_{i+1})]$ be taken, for the interval (x_i, x_{i+1}).

An Improvement to Chiang's Approach

Keyfitz and Frauenthal (1975) suggested an improvement to Chiang's (1968) suggestion. The improvement consists in applying a multiplier to Chiang's estimate using a somewhat new notation, $M_\alpha(x, x + n)/M_+(x, x + n)$. The multiplier is

$$1 + \frac{1}{24}\left[\frac{M_\alpha(x+n, x+2n) + M_\alpha(x-n, x)}{M_\alpha(x, x + n)} - \frac{M_+(x+n, x+2n) + M_+(x-n, x)}{M_+(x, x + n)}\right]$$

(8.51)

Their empirical examination of the magnitude of the correction, however, led them to conclude that "it is genuinely indifferent whether the correction is made or not. Nonetheless cases will arise where within an age group one cause is declining and others are rising, and then the correction will bring improvement" (Keyfitz and Frauenthal, 1975, p. 898).

6 Follow-Up Studies

Ungrouped Data

Suppose n individuals are followed until death or censoring. For each individual who dies the time until death and the cause of death are recorded. For each censored case the time until censoring is recorded. As before let $\mu(t, \alpha)$ denote the cause-specific force of mortality for cause C_α, and let the total force of mortality for all causes collapsed be $\mu(t, +)$. Let us define

$$H(t, \alpha) = \int_0^t \mu(s, \alpha)\, ds$$

(8.52)

$$H(t, +) = \int_0^t \mu(s, +)\, ds$$

(8.53)

$$G(t, \alpha) = \exp[-H(t, \alpha)]$$

(8.54)

and

$$G(t, +) = \exp[-H(t, +)]$$

(8.55)

Clearly,

$$G(t, +) = \prod_\alpha G(t, \alpha)$$

(8.56)

Recall that $\mu(t, \alpha)$ is not in general equal to the cause-specific force of mortality obtained from the marginal distribution corresponding to cause C_α, and therefore $G(t, \alpha)$ does not in general have any survival function interpretation. However, when the potential failure times are independently distributed, $G(t, \alpha)$ gives the net survival function.

One uses maximum likelihood procedures for statistical inference under the setup. (The reader is advised to review Chapter 10.) The likelihood on the data is the product of contributions from the censored and uncensored cases. The contribution from a censored case is of the form $G(t, +)$, where t is the time until censoring, while the contribution from an uncensored case is of the form $\mu(t, \alpha)G(t, \alpha)$, where t is the time until death and α the cause of death. If n_i is the number of individuals at risk just prior to t_i and $d_{\alpha i}$ is the number of individuals dying from cause α at t_i, then the maximum likelihood estimate of $G(t, \alpha)$ is given by

$$\hat{G}(t, \alpha) = \prod_{t: t_i < t} \frac{n_i - d_{\alpha i}}{n_i} \tag{8.57}$$

The marginal survival function can be estimated by the PL method, from the pooled data (all causes combined), using the following formula:

$$\hat{S}(t) = \prod_{t: t_i < t} \frac{n_i - d_i}{n_i} \tag{8.58}$$

where d_i is the number of deaths (from all causes combined) at t_i. To estimate the cause-specific survival function, one uses the formula

$$\hat{S}(t, \alpha) = \prod_{t: t_i < t} \frac{d_{\alpha i}}{n_i} \hat{S}(t_i) \tag{8.59}$$

and to estimate the expected proportion of death from cause α, the following formula may be used:

$$\hat{\pi}_\alpha = \hat{S}(0, \alpha) \Big/ \sum_\alpha \hat{S}(0, \alpha) \tag{8.60}$$

(When there is no censoring, the sum $\sum_\alpha \hat{S}(0, \alpha)$ will be unity.)

Grouped Data

Suppose the time axis is divided into I fixed, disjoint intervals and the ith interval is $[x_i, x_{i+1})$, $i = 0, \ldots, (I - 1)$. Let N_i be the number of individuals surviving until the beginning of the ith interval, and let $D_{i\alpha}$ be the number of deaths from cause C_α.

When There is No Censoring

First let us consider the case of complete data. Writing

$$D_{i+} = D_{i1} + \cdots + D_{ik} \tag{8.61}$$

we notice that

$$N_{i+1} = N_i - D_{i+} \tag{8.62}$$

If $q(i, +)$ and $q(i, \alpha)$ stand for the conditional probabilities of dying in the ith interval, given survival until the beginning of the interval, from any cause and cause C_α, respectively, then given N_i, the $D_{i\alpha}$'s and N_{i+1} are jointly distributed as a multinomial, with probability function proportional to

$$\left[1 - \sum_\alpha q(i, \alpha)\right]^{N_{i+1}} \prod_\alpha [q(i, \alpha)]^{D_{i\alpha}} \tag{8.63}$$

It follows that the maximum likelihood estimates are

$$\hat{q}(i, \alpha) = D_{i\alpha}/N_i, \qquad i = 1, \ldots, k, \tag{8.64}$$

and

$$\hat{q}(i, +) = D_{i+}/N_i \tag{8.65}$$

From (8.64) and (8.65) we have

$$\hat{q}(i, \alpha) = [D_{i\alpha}/D_{i+}]\hat{q}(i, +) \tag{8.66}$$

The probability of surviving from the beginning of the ith interval until that of the next interval is $1 - q(i, +)$, the maximum likelihood estimate of which is N_{i+1}/N_i. The probability, say P_i, of surviving until the beginning of the ith interval is the product

$$\prod_{j=0}^{i-1} [1 - q(j, +)] \tag{8.67}$$

an estimate of which is N_i/N_0, where N_0 is the number of individuals under observation at the beginning of the first interval. Also the "survival" function, say $P_{i\alpha}$, for cause C_α in the presence of all causes, is estimated by

$$\sum_{j=i}^{I-1} \hat{q}(j, \alpha)\hat{P}_i \tag{8.68}$$

which is the same as

$$\frac{1}{N} \sum_{j=i}^{I-1} D_{i\alpha} \tag{8.69}$$

and the conditional probability of dying from cause C_α, given survival to the beginning of the ith interval is estimated as the ratio

$$\hat{P}_{i\alpha}/\hat{P}_i \tag{8.70}$$

When There is Censoring

Let us now turn to the case of incomplete data, that is, data involving censoring. In addition to the notation introduced above, let W_i be the number of subjects censored in the ith interval. In this situation, we use the approximation

$$\hat{q}(i, +) \simeq D_{i+}/(N_i - \tfrac{1}{2}W_i) \tag{8.71}$$

and

$$\hat{q}(i, \alpha) \simeq D_{i\alpha}/(N_i - \tfrac{1}{2}W_i) \tag{8.72}$$

Once $\hat{q}(i, +)$ and $\hat{q}(i, \alpha)$ are on hand, we proceed with the estimation of other quantities in the manner shown above for the uncensored case, remembering that instead of N_i in the uncensored case we work with $N_i^* = N_i - \tfrac{1}{2}W_i$, which is sometimes called the effective number surviving until the beginning of the ith interval.

Problems and Complements

1. The term "competing risks" is applied to problems in which an object is exposed to two or more causes of failure. Such problems crop up in many contexts. Thus, individual humans are subject to multiple causes of death. Death and disability are competing risks for purposes of calculations of premiums for disability insurance. If failure of any subsystem (part) may result in the failure of the total system (whole), then the failure of the latter can be described in terms of failures of particular subsystems. [Think of a physical system composed of two or more parts, and assume that the system fails if any of its parts fails. Assume also that an observer can determine which part of the system failed first.] For example consider an actuary calculating premiums for joint life insurance for married couples. The actuary has to consider the competing risks of death of the husband and the wife. In calculating the benefit of reducing exposure to cancer causing agents, one takes into account causes of death other than cancer. Although the analogy between causes of death of humans and causes of failure of physical systems is crude, scholars have found it useful to draw on it when developing theories and analysis schemes. The actuarial literature on competing risks has been reviewed by Seal (1977). The statistical literature has been reviewed by Chiang (1968), David and Moeschberger (1978), Gail (1975), and Prentice *et al.* (1978), among others.

2. The term "cause-specific hazard functions" is used by Prentice *et al.* (1978) for what has been called cause-specific forces of mortality in this chapter. These are the same as the "force of transition" functions in Aalen's (1976) exposition of the competing risk problem as a Markov process. Makeham (1860) used the term "decremental forces" for these.

3. Suppose $\mu(x, \alpha)$ is the cause-specific force of mortality and $\mu(x, +)$ the corresponding force of mortality when cause of death is ignored. Then the relationship $\mu(x, +) = \Sigma_\alpha \mu(x, \alpha)$ holds irrespective of whether the causes of death are independent.

4. The conditional probability of dying of cause α in the interval $(\tau_{i-1}, \tau_i]$ given survival until τ_{i-1} is termed "crude" probability by Chiang (1968). These conditional probabilities are the basic parameters of the multiple-decrement life table.

5. Different authors use different notation for the parameters of the multiple-decrement life table. Using the notation $Q(j; i)$ for the conditional probability of dying of cause j in the ith interval given survival until the ith interval, where the positive time axis is partitioned by cut points at τ_i, $i = 0, 1, \ldots$, with $\tau_0 = 0$, what is the expression for the conditional probability of surviving the interval i? [ANSWER: $\rho_i = 1 - \Sigma_j Q(j; i)$, where the summation is over causes of death, i.e., over j.] What does the product $\rho_1 \cdots \rho_i$ stand for? [ANSWER: The probability of surviving beyond the beginning of the ith interval, i.e., beyond τ_i.] What interpretation would you give $\Sigma_{i=1}^k \rho_{i-1} Q(j; i)$. [ANSWER: The probability of dying of cause j before age τ_k.] If d_{ji} stands for the number of deaths of cause j in the ith interval and n_i for the number at risk as of the beginning of that interval, what is the maximum likelihood estimate of $Q(j; i)$? [ANSWER: d_{ji}/n_i.] Under the setup just described, what is the maximum likelihood estimate of ρ_i? [ANSWER: s_i/n_i, where $s_i = n_i - \Sigma_j d_{ji}$.]

6. Chiang (1968) has defined the "net" probability q_{ji} as the conditional probability of dying in the ith interval given survival until the beginning of that interval, when the only risk operating is j. He has defined $Q_{ji,\delta}$ as the corresponding "partial crude" probability of dying of risk j when risk δ has been eliminated. Prentice et al. (1978) call such quantities "nonobservable" on the basis of the convention that only functions of the cause-specific forces of mortality corresponding to the original observations when all risks are acting are "observable." Quantities such as the ones Chiang has introduced cannot be estimated without introducing additional assumptions, i.e., without the aid of an assumed model. In the text each individual was assumed to have k failure times $X_1, \ldots, X_k$, of which only $X = \min(T_1, \ldots, T_k)$ and the risk j are observed. Such a model is sometimes called the "latent-failure time model." To use this model to compute "nonobservable" probabilities such as the ones mentioned above, a joint survival function $S(X_1, \ldots, X_k)$ is introduced. When the partial derivatives $\partial \ln S(x_1, \ldots, x_k)/\partial x_j$ evaluated at $x_1 = \cdots = x_k = x$ exist, the negative of these partial derivatives (evaluated at $x_1 = \cdots = x_k = x$) are called cause-specific forces of mortality. If $\lambda(x)$ stands for the limit as Δx tends to zero of $(1/\Delta x) \Pr[x \le X < x + \Delta x | X \ge x]$, show that $\lambda(x)$ satisfies the relation $\lambda(x) = -d \ln S(x)/dx$.

7. Suppose that under the setup described in Exercise 6, $X_1, \ldots, X_k$ are independent with marginal survival distributions $S_j(x) = \Pr[X > x]$ and corresponding marginal forces of mortality $\lambda_j(x) = -d \ln S_j(x)/dx$. Show that under independence $S(x_1, \ldots, x_k) = \Pi S_j(x)$, and hence that cause-specific force of mortality for cause j is equal to the corresponding marginal force of mortality. Is the converse necessarily true? [ANSWER: No (see Gail, 1975).]

8. Suppose there are only three causes of death. The net probability of surviving until age x when cause C_1 alone is acting is $S(x, 0, 0) = exp[-\int_0^x \lambda_1(u) \, du]$. Also, the partial crude probability of dying of cause C_3 in $[0, x]$ with cause C_1 eliminated is $\int_0^x \lambda_3(v) S_2(v) S_3(v) \, dv$. Under the assumption of independence, show that the latter is equal to $\int_0^x \lambda_3(v) S(0, v, v) \, dv$. In this calculation the net survival distribution for cause C_1 was taken to be the marginal distribution $S_1(x) = S(x, 0, 0)$. A generalization of this is to assume that the effect of eliminating cause C_j is to set the corresponding argument in $S(x_1, \ldots, x_k)$ equal to zero. Thus, not only do we assume that the structure of $S(x_1, \ldots, x_k)$ is known but also that the effect of eliminating a subset of causes of death can be expressed in terms of $S(x_1, \ldots, x_k)$. Gail (1975) has discussed the implications of this approach to the estimation of various "observable" and "nonobservable" probabilities.

9. Elandt–Johnson (1976) has suggested that the net survival distribution for cause C_1 be defined as the limit of $\Pr[X_1 > x | X_2 = x_2, \ldots, X_k = x_k]$ as $x_2, x_3, \ldots$, and x_k all tend to infinity. If $X_1, X_2, \ldots, X_k$ are independent, the net survival distribution thus defined and $S_1(x) = S(x, 0, \ldots, 0)$ are equivalent. Thus, under independence the net survival distributions may be

taken to be the corresponding marginal distributions of $S(x_1, \ldots, x_k)$, and, more generally, the effect of eliminating cause C_j may be modeled by setting the corresponding argument in $S(x_1, \ldots, x_k)$ to zero. When independence cannot be assumed, however, estimates of nonobservable probabilities depend on the hypothesized structure of the joint survival distribution.

10. Consider the random variables $X_1, \ldots, X_k$ with a survival distribution $S(x_1, \ldots, x_k)$. Let $\mu(x, \alpha)$, $\alpha = 1, \ldots, k$ be the cause-specific forces of mortality corresponding to $S(x_1, \ldots, x_k)$. Now let X_i^*, $i = 1, \ldots, k$ be k random variables such that $S^*(x_1, \ldots, x_k) = \Pi_\alpha \exp[-\int_0^x \mu(u, \alpha) \, du]$. It is clear that $S^*(x_1, \ldots, x_k)$ thus defined has the same cause-specific forces of mortality as the original $S(x_1, \ldots, x_k)$. This implies that even if the data are so complete that we can specify the cause-specific forces of mortality exactly, they do not define $S(x_1, \ldots, x_k)$ uniquely. Tsiatis (1975) has called this problem that of *nonidentifiability*. One of the implications of nonidentifiability is that the structure of $S(x_1, \ldots, x_k)$ cannot be empirically verified. Another implication is that one can estimate the cause-specific forces of mortality corresponding to any survival distribution $S(x_1, \ldots, x_k)$ under the hypothesis that the survival distribution is $S^*(x_1, \ldots, x_k)$ defined above.

11. Aalen (1976) has viewed the classical competing risk problem as a continuous-time Markov process, with one alive state $\alpha = 0$ and a multiplicity of absorbing death states, the only permissible transitions being from the alive state to one or another of the absorbing states. It is also assumed that the elimination of a cause of death merely nullifies the corresponding cause-specific force of mortality. Show that Aalen's modeling is equivalent to the approach described in this chapter in terms of independent latent times to death (failure).

12. Consider the estimate $D_{i\alpha}/N_i^*$ of $q(i, \alpha)$. The variances and covariances of these estimates can be estimated by viewing them as estimates of multinomial parameters. Thus we estimate the variance of $\hat{q}(i, \alpha)$ as

$$\hat{q}(i, \alpha)[1 - \hat{q}(i, \alpha)]/N_i^*$$

and the covariance of $\hat{q}(i, \alpha)$ and $\hat{q}(i, \alpha')$ as

$$-\hat{q}(i, \alpha)\hat{q}(i, \alpha')/N_i^*$$

Suppose we want to estimate the cumulative probability that an individual will die of cause C_α before age x_{i+1}. An estimate of this probability is

$$\hat{q}(0, \alpha) + \hat{p}_0\hat{q}(1, \alpha) + \cdots + \hat{p}_0\hat{p}_1 \cdots \hat{p}_{i-1}\hat{q}(i, \alpha)$$

where $\hat{p}_i = 1 - \Sigma \, \hat{q}(i, \alpha)$. What we have here is a complicated function of a number of statistics (the $\hat{q}$'s) as an estimate of a probability of interest. The variance of such functions of statistics can be approximately expressed in terms of the variances and covariances of the individual statistics (i.e., the $\hat{q}$'s) using what is known as the delta method or the method of statistical differentials, a brief description of which follows. Let $g(X)$ be a function of a random variable X. Let the expected value of X be θ. Then using Taylor series expansion we have

$$g(X) = g(\theta) + (X - \theta)g'(\theta) + \tfrac{1}{2}(x - \theta)^2 g''(\theta) + \cdots$$

Taking expectations, and noticing that the expectation of $(X - \theta)$ is zero, we obtain, ignoring terms involving third and higher powers of $(X - \theta)$, the approximate formula

$$\text{expected value of } g(X) \simeq g(\theta) + \tfrac{1}{2}g''(\theta) \, \text{Var}(X)$$

[Very often only the linear part of the Taylor expansion is retained, which leads to the approximation of the expected value of $g(X)$ by $g(\theta)$.] Subtracting the approximation closer to the given expected value of $g(X)$ from $g(X)$, squaring the deviation, and then taking the expectation we get $[g'(\theta)]^2$ var(X) as an approximation to the variance of $g(X)$. When we have a function of two or more random variables, we use essentially the same approach, recognizing, of course, that the Taylor expansion then involves partial derivatives. The formula thus derived, for the case of two random variables is

$$\text{var}[g(X_1, X_2)] \simeq (\partial g/\partial \theta_1)^2 \text{var}(X_1) + (\partial g/\partial \theta_2)^2 \text{var}(X_2)$$
$$+ 2(\partial g/\partial \theta_1)(\partial g/\partial \theta_2) \text{cov}(X_1, X_2)$$

where θ_i is the expected value of X_i, $i = 1, 2$. Using this approach, the variance of estimated probability that an individual will die of cause C_α before age x_{i+1} can be shown to be

$$\sum_{h=0}^{i-1} A_h^2 \text{var}(\hat{p}_h) + \sum_{h=0}^{i} B_h^2 \text{var}[\hat{q}(h, \alpha)] + 2 \sum_{h=0}^{i} A_h B_h \text{cov}[\hat{q}(h, \alpha), \hat{p}_h)]$$

where

$$A_h = \frac{1}{p_h} \sum_{j=h}^{i-1} (p_0 \times p_1 \times \cdots \times p_j) q(j, \alpha)$$

$$B_h = p_0 \times p_1 \times \cdots \times p_{h-1}, \quad \text{with} \quad B_0 = 1$$

$$\text{cov}[\hat{q}(h, \alpha), \hat{p}_h] = -p_h q(h, \alpha)/N_h^*$$

$$\text{var}(\hat{p}_h) = p_h q_h/N_h^*,$$

and

$$\text{var}[q(h, \alpha)] = [q(h, \alpha)\{1 - q(h, \alpha)\}]/N_h^*.$$

Bibliographic Notes

Some historical material concerning the development of competing risk theory is available in David and Moeschberger (1978) and Birnbaum (1979). Details regarding the equivalence theorem are presented in Cox (1962), Tsiatis (1975), Elandt-Johnson and Johnson (1980), and Langberg *et al.* (1978), among others. The problem of nonidentifiability has been examined theoretically by Nadas (1971) and by Basu and Gosh (1978). Other pertinent references in connection with nonidentifiability and elimination of causes of death include Elandt-Johnson (1976) and Kimball (1958, 1969). Gail's (1975) paper is an often-quoted critique of competing risk analysis. Some new perspectives have been added by Manton and Stallard through their work on cause-of-death analyses (see Manton and Stallard 1984, and the references cited therein; also see Prentice *et al.*, 1978; Lagakos *et al.*, 1978; Fleming, 1978a,b; Fix and Neyman, 1951; Chiang, 1968; Kalbfleisch and Prentice, 1980; and Miller and Hickman, 1983).

Chapter 9 | Multistate Life Tables

As we have seen, the ordinary life table deals with the time pattern of attrition of a well-defined population as the members of the population move from a transient state, such as being alive, to an absorbing state, such as being dead. When two or more absorbing states are involved, we use the multiple-decrement life table, and when two or more transient states are involved, as in the analysis of labor-force participation or nuptiality, we use the increment–decrement, or multistate, life table. This chapter is devoted to this last type of life table.

Among the early expositions of the multistate life table are Fix and Neyman (1951), Chiang (1964), Sverdrup (1965), Jordan (1967), and Hoem (1970). More recent works on the subject include those of Rogers (1973), Rogers and Ledent (1976), Hoem and Fong (1976), Krishnamoorthy (1979b), Schoen and Land (1979), Ledent (1980), Hoem and Jensen (1982), Mode (1982), Willekens (1982), and Nour and Suchindran (1983, 1984a,b, 1985).

1 Theory

Suppose individuals independently move among n *states*. As an example, in the case of nuptiality the states may be (1) "never married," (2) "married," (3) "divorced," (4) "widowed," and (5) "dead." Together, the states constitute what is known as the *state space*. The probabilistic structure of the multi-

137

state life table is that of a continuous-time nonhomogeneous Markov process with finite state space. For convenience of reference we number the states from 1 to n. Let $X(t)$ be the state occupied at time t by a randomly chosen individual. For $0 \le s \le t$, let $q_{ij}(s, t)$ be the probability that $X(t) = j$ given $X(s) = i$. Let the symbol H refer to the state space, and H_1 and H_2, respectively, refer to the subspaces constituted by the transient and absorbing states. Let the numbers of states in H_1 and H_2 be n_1 and n_2, respectively. Since $X(t)$ changes continuously with time, there is no basic time unit that can be used to describe what we ordinarily call the "next move." This necessitates the introduction of the following concepts. We assume that the following limits exist:

$$r_{ij}(s) = \lim_{u \to 0} q_{ij}(s, s + u)/u, \qquad i \ne j, \quad i, j \in H \qquad (9.1)$$

The quantity $r_{ij}(s)$ is nonnegative, but it is not necessarily between 0 and 1. It is called the intensity (force) of transition from state i to state j at time s. We introduce for convenience $r_{ii}(s)$ such that

$$\sum_j r_{ij}(s) = 0 \qquad (9.2)$$

It is easy to see that

$$r_{ii}(s) = \lim_{u \to 0} \frac{[q_{ii}(s, s + u) - 1]}{u} \qquad (9.3)$$

The quantity, $-r_{ii}(s)$, is nonnegative; it is sometimes referred to as the *intensity* (*force*) *of passage*, because it concerns the passage from state i to one or another state different from i. Let us use the symbol $\mathbf{Q}(s, t)$ for the matrix whose (i, j) element is $q_{ij}(s, t)$, and the symbol $\mathbf{R}(s)$ for the matrix whose (i, j) element is $r_{ij}(s)$. Each row of $\mathbf{Q}$ sums to unity, whereas each row of $\mathbf{R}$ sums to zero. Also

$$\lim_{u \to 0} \frac{\mathbf{Q}(t, t + u) - \mathbf{I}}{u} = \mathbf{R}(t) \qquad (9.4)$$

It is useful to partition $\mathbf{Q}(s, t)$ so that the entries in the top n_1 rows and the first n_1 columns pertain to the transient states:

$$\mathbf{Q}(s, t) = \begin{bmatrix} \mathbf{Q}_{11}(s, t) & \mathbf{Q}_{12}(s, t) \\ \mathbf{Q}_{21}(s, t) & \mathbf{Q}_{22}(s, t) \end{bmatrix} \qquad (9.5)$$

where the bottom left matrix $\mathbf{Q}_{21}(s, t)$ is a null matrix, i.e., one with all of its elements zero, there being no transition from absorbing states to any transition state, and the bottom right matrix $Q_{22}(s, t)$ is an identity matrix, since by definition once an absorbing state is entered into, there is no escape to any state from it. The corresponding partitioning of $R(s)$ is

$$R(s) = \begin{bmatrix} R_{11}(s) & R_{12}(s) \\ 0 & 0 \end{bmatrix} \tag{9.6}$$

where 0 denotes a null matrix.

Let us now consider three time points $s \leq t \leq (t + u)$. The Markovian assumption implies the following relationship between $Q(s, t)$, $Q(t, t + u)$, and $Q(s, t + u)$:

$$Q(s, t + u) = Q(s, t) \times Q(t, t + u) \tag{9.7}$$

If we subtract $Q(s, t)$ from both sides of (9.7) we get

$$Q(s, t + u) - Q(s, t) = Q(s, t)[Q(t, t + u) - I] \tag{9.8}$$

Now if we divide both sides of (9.8) by u and then take the limit as u tends to 0, we get, by virtue of (9.4),

$$\partial Q(s, t)/\partial t = Q(s, t) \times R(t) \tag{9.9}$$

Similarly, by noting that, analogous to (9.7), we have

$$Q(s, t) = Q(s, s + v) \times Q(s + v, t) \tag{9.10}$$

we obtain [by subtracting $Q(s + v, t)$ from both sides, then dividing both sides by v, and finally taking limit as v tends to 0]

$$\partial Q(s, t)/\partial s = -R(s) \times Q(s, t) \tag{9.11}$$

Equations (9.10) and (9.11), respectively, are known as the "forward" and "backward" Chapman–Kolmogorov systems. We can obtain an equation system similar to (9.10) for the row vector $P(t)$, whose ith element denotes the probability that a randomly selected individual is in state i at time t. First, notice that by virtue of the Markovian assumption,

$$P(t) = P(s) \times Q(s, t) \tag{9.12}$$

Differentiation gives

$$\partial P(t)/\partial t = P(s)\, \partial Q(s, t)/\partial t \tag{9.13}$$

the right-hand side of which is equal to $P(s) \times Q(s, t) \times R(t)$, as premultiplication of (9.9) with $P(s)$ implies, which in turn is equal to $P(t) \times R(t)$ by virtue of (9.12), thus giving

$$\partial P(t)/\partial t = P(t)R(t) \tag{9.14}$$

Summary Measures

In the ordinary life table, an often used summary measure is the expected remaining life time at age x, which, it may be recalled, is the person-years lived, on average, beyond age x by those surviving to age x. One could compute similarly the person-years lived on average during a fixed time interval, beyond age x, by those surviving to age x. [The former, of course, is a particular case of the latter when the fixed period coincides with the entire lifespan beyond age x.] In the multistate life table, we similarly use as a summary measure the expected length of sojourn time in state j between times s and t, given occupancy of state i at time s. Let us imagine observing a process that occupies a nonabsorbing state i at time s. After staying there for a random duration, it moves to another state. If the latter is a nonabsorbing state, the process stays there for a random duration and then moves to another state. This continues until the observation period ends or an absorbing state is reached. The expected sojourn time in state i is the expected value of the sum of the random durations of occupancy of state i in this evolution. In nuptiality processes, with states "never married," "married," "divorced," "widowed," and "dead," we speak of, for example, the expected length of time a randomly chosen person remains in the divorced state during the age interval 20 to 30, given that the person was married as of age 20. Notice that the number of divorces (the number of visits to the divorced state) during this age interval is itself a random variable, taking integer values 0, 1, 2, etc.

Let $E(s, t)$ be a matrix whose (i, j) element, $e_{ij}(s, t)$, is the expected sojourn time in state j between times s and t, conditional on occupancy of state i at time s. We can show that

$$E(s, t) = \int_s^t Q(s, u) \, du \tag{9.15}$$

Let $I_{ij}(s, u)$ be an indicator variable that takes values 1 or 0 depending upon whether the process that occupies state i at time s is in state j at time u. The sojourn time mentioned above is then the point-wise integral

$$\int_s^t I_{ij}(s, u) \, du$$

whose expectation is

$$e_{ij}(s, t) = \int_s^t q_{ij}(s, u) \, du$$

which gives the (i, j) element of $E(s, t)$ in (9.15). It may be noted in passing that if we treat $E(s, t)$ in (9.15) as a function of s, keeping t fixed, we obtain on differentiation

$$\partial \mathbf{E}(s, t)/\partial s = -\mathbf{R}(s) \times \mathbf{E}(s, t) - \mathbf{I} \tag{9.16}$$

and, conversely, (9.15) can be obtained by solving (9.16), subject to the initial condition $\mathbf{E}(t, t) = 0$.

Another summary measure frequently used in multistate life tables involves the number of visits made to a transient state during a fixed time interval. Let $m_{ij}(s, t)$ denote the expected number of visits to state j between times s and t, given occupancy of state i at time s. It can be shown that if $\mathbf{M}(s, t)$ has $m_{ij}(s, t)$ for its (i, j) element, then

$$\mathbf{M}(s, t) = \int_s^t \mathbf{Q}(s, u)\mathbf{B}(u) \, du \tag{9.17}$$

where $\mathbf{B}(u)$ is obtained from $\mathbf{R}(u)$ by replacing the latter's diagonal entries with zeros. The expression just given for $\mathbf{M}(s, t)$ satisfies the differential equation

$$\partial \mathbf{M}(s, t)/\partial s = -\mathbf{R}(s) \times \mathbf{M}(s, t) - \mathbf{B}(s) \tag{9.18}$$

Conversely, (9.17) can be obtained by solving (9.18) subject to the initial condition $\mathbf{M}(t, t) = \mathbf{I}$.

There are of course several other summary measures one can think of. But from what has already been said, it should be clear that the matrix of transition probabilities is fundamental—other functions of the multistate life table can be expressed in terms of this fundamental matrix.

Estimation

We have noted that an important property of a continuous-time nonhomogeneous Markov process is that the transition matrix function $\mathbf{Q}(s, t)$ satisfies

$$\partial \mathbf{Q}(s, t)/\partial t = \mathbf{Q}(s, t) \times \mathbf{R}(t), \qquad \mathbf{Q}(t, t) = \mathbf{I} \tag{9.19}$$

For a given $\mathbf{R}(t)$, there exists a $\mathbf{Q}(s, t)$, which can be expressed as

$$\mathbf{Q}(s, t) = \mathbf{I} + \int_s^t \mathbf{Q}(s, u) \times \mathbf{R}(u) \, du \tag{9.20}$$

satisfying (9.19). From (9.20) we have

$$\mathbf{Q}(t, t + h) = \mathbf{I} + \int_t^{t+h} \mathbf{Q}(t, u) \times \mathbf{R}(u) \, du \tag{9.21}$$

By applying the mean value theorem, we obtain from (9.21)

$$\mathbf{Q}(t, t + h) = \mathbf{I} + \left[\int_0^h \mathbf{Q}(t, t + u) \, du \right] \mathbf{R}^*(t, t + h) \tag{9.22}$$

where $\mathbf{R}^*(t, t + h)$ has $r_{ij}^*(t, t + h)$ as its (i, j) element, which lies between the minimum and the maximum values of $r_{ij}(t + u)$ over $0 \le u \le h$. From (9.21) we obtain

$$\mathbf{R}^*(t, t + h) = \left[\int_0^h \mathbf{Q}(t, t + u) \, du \right]^{-1} [\mathbf{Q}(t, t + h) - \mathbf{I}] \qquad (9.23)$$

Using the Markovian property of $\mathbf{Q}$, if we replace in (9.23) $\mathbf{Q}(t, t + u)$ with $[\mathbf{Q}(0, t)]^{-1} \times \mathbf{Q}(0, t + u)$ and $\mathbf{Q}(t, t + h)$ with $[\mathbf{Q}(0, t)]^{-1} \mathbf{Q}(0, t + u)$, we obtain, remembering that $\mathbf{I}$ can be replaced with $[\mathbf{Q}(0, t)]^{-1} \mathbf{Q}(0, t)$,

$$\mathbf{R}^*(t, t + h) = \left[\int_0^h \mathbf{Q}(0, t + u) \, du \right]^{-1} [\mathbf{Q}(0, t + u) - \mathbf{Q}(0, t)] \qquad (9.24)$$

which is the usual matrix of interval (age)-specific occurrence–exposure rates of transition. In practice for estimation purposes we equate the occurrence–exposure rate observed in a population to the corresponding life table value. Equating life table rates to observed occurrence–exposure rates is valid only if sectional stationarity prevails, i.e., within each age interval and with each life status of the model, the observed population is stationary. For alternative assumptions see, e.g., Keyfitz (1970) and Keyfitz et al. (1972). The alternative assumptions in use do not seem to make any important difference in numerical applications (Hoem and Jensen, 1982). Following this procedure, we obtain as an estimate of $r_{ij}^*(t, t + u), i \ne j$

$$N_{ij}(t, h) \bigg/ \int_0^h N_i(t + u) \, du \qquad (9.25)$$

where $N_{ij}(t, h)$ stands for the number of transitions from state i to state j over the interval t to $t + h$, and $N_i(t + u)$ represents the number of individuals in state i at $t + u, 0 \le u \le h$. Once $r_{ij}^*(t, t + h)$ for $i \ne j$ have been estimated, we obtain $r_{ii}^*(t, t + h)$ using the relation $\Sigma_j r_{ij}^*(t, t + h) = 0$.

Let us assume that the rate of transition from state i to state j is the same over all possible subintervals of $[t, t + h]$ for all (i, j) combinations, $i, j \in H$. This is a reasonable assumption if we choose h to suit the circumstances. Given this assumption we can treat $R^*(t, t + u)$ as a constant in the interval $[t, t + h]$, and, consequently, differentiation of (9.22) gives

$$\partial \mathbf{Q}(t, t + u)/\partial u = \mathbf{Q}(t, t + u)\mathbf{R}^*(t, t + u) \qquad (9.26)$$

or, equivalently,

$$\partial \mathbf{Q}_{11}(t, t + u)/\partial u = \mathbf{Q}_{11}(t, t + u) \times \mathbf{R}_{11}^*(t, t + u) \qquad (9.27)$$

and

$$\partial \mathbf{Q}_{12}(t, t + u)/\partial u = \mathbf{Q}_{11}(t, t + u) \times \mathbf{R}_{12}^*(t, t + u) \qquad (9.28)$$

Computation of $\mathbf{Q}(s, t)$

Equation (9.27) has the solution

$$\mathbf{Q}_{11}(t, t + u) = \exp[u \times \mathbf{R}^*_{11}(t, t + u)] \qquad (9.29)$$

given the initial condition $\mathbf{Q}_{11}(t, t) = \mathbf{I}$. Solving (9.28) after inserting the solution (9.29), gives

$$\mathbf{Q}_{12}(t, t+u) = [\mathbf{Q}_{11}(t, t+u) - \mathbf{I}][\mathbf{R}^*_{11}(t, t+u)]^{-1}\mathbf{R}^*_{12}(t, t+u) \qquad 0 \leq u \leq h \qquad (9.30)$$

provided that $\mathbf{R}^*_{11}(t, t + u)$ is nonsingular. Once we estimate, $\mathbf{Q}_{11}$ and $\mathbf{Q}_{12}$, we can complete $\mathbf{Q}$ by concatenating the partitions together, remembering that $\mathbf{Q}_{21}$ is a null matrix, and $\mathbf{Q}_{22}$ is an identity matrix. To estimate $\mathbf{Q}(s, t)$, we partition the interval $[s, t]$ into as many subintervals as are necessary in order to permit the application of (9.29) and (9.30) to each subinterval. If $s = t_0 \leq t_1 \leq t_2 \leq \cdots \leq t_n = t$ is one such partitioning, we obtain the formula

$$\mathbf{Q}(s, t) = \mathbf{Q}(t_0, t_n) = \mathbf{Q}(t_0, t_1) \times \mathbf{Q}(t_1, t_2) \times \cdots \times \mathbf{Q}(t_{n-1}, t_n) \qquad (9.31)$$

Computation of $\mathbf{E}(s, t)$

From (9.15) we have for $0 \leq u \leq h$

$$\mathbf{E}_{11}(t, t + u) = \int_0^u \mathbf{Q}_{11}(t, t + v)\, dv \qquad (9.32)$$

$$\mathbf{E}_{12}(t, t + u) = \int_0^u \mathbf{Q}_{12}(t, t + v)\, dv \qquad (9.33)$$

It may be noted in passing that the elements of $\mathbf{E}_{12}$ may be thought of as the expected loss of time due to transition to an absorbing state. By inserting in (9.32) the solution for $\mathbf{Q}_{11}(t, t + u)$ from (9.29) and then integrating, we get

$$\mathbf{E}_{11}(t, t + u) = [\mathbf{Q}_{11}(t, t + u) - \mathbf{I}][\mathbf{R}^*_{11}(t, t + u)]^{-1} \qquad (9.34)$$

and similarly from (9.30) and (9.33)

$$\mathbf{E}_{12}(t, t + u) = [\mathbf{E}_{11}(t, t + u) - \mathbf{I}][\mathbf{R}^*_{11}(t, t + u)]^{-1}\mathbf{R}^*_{12}(t, t + u) \qquad (9.35)$$

By concatenating $\mathbf{E}_{11}$, $\mathbf{E}_{12}$, $\mathbf{E}_{21}$, and $\mathbf{E}_{22}$, the last two being null matrices, we obtain the full matrix $\mathbf{E}(t, t + u)$. To construct $\mathbf{E}(s, t)$, we note that

$$\mathbf{E}(s, t) = \int_s^t \mathbf{Q}(s, u)\, du$$

$$= \sum_{i=0}^{n-1} \int_{t_i}^{t_{i+1}} \mathbf{Q}(s, u), \qquad s = t_0 \leq t_1 \leq \cdots \leq t_n = t \qquad (9.36)$$

Writing h_i for $t_{i+1} - t_i$

$$\int_{t_i}^{t_{i+1}} \mathbf{Q}(s, u) \, du = \int_{t_i}^{t_i + h_i} \mathbf{Q}(s, u) \, du = \int_0^{h_i} \mathbf{Q}(s, t_i + u) \, du$$

$$= \mathbf{Q}(s, t_i) \int_0^{h_i} \mathbf{Q}(t_i, t_i + u) \, du$$

$$= \mathbf{Q}(s, t_i) \mathbf{E}(t_i, t_{i+1}) \tag{9.37}$$

Thus

$$\mathbf{E}(s, t) = \sum_{i=0}^{n-1} \mathbf{Q}(s, t_i) \mathbf{E}(t_i, t_{i+1}) \tag{9.38}$$

Computation of $\mathbf{M}(s, t)$

Partitioning $\mathbf{M}(s, t)$ in a fashion corresponding to that of $\mathbf{Q}(s, t)$ and $\mathbf{E}(s, t)$ introduced above, we notice that

$$\mathbf{M}_{11}(s, t) = \int_s^t \mathbf{Q}_{11}(s, u) \mathbf{B}_{11}(s, u) \, du \tag{9.39}$$

$$\mathbf{M}_{12}(s, t) = \mathbf{Q}_{12}(s, t) \tag{9.40}$$

and $\mathbf{M}_{21}(s, t)$ and $\mathbf{M}_{22}(s, t)$ are null matrices. [Equation (9.40) follows from the fact that the elements of $\mathbf{M}_{12}$ represent the expected number of visits from various transient states to different absorbing states over a given interval of time. By definition, only one such visit is possible for an individual, there being no escape from any absorbing state. The probability of making a visit to an absorbing state is the corresponding element of the appropriate $\mathbf{Q}$ matrix.]

For the partitioning, $s = t_0 \leq t_1 \leq \cdots \leq t_n = t$,

$$\mathbf{M}_{11}(s, t) = \sum_{i=0}^{n-1} \int_{t_i}^{t_{i+1}} \mathbf{Q}_{11}(s, u) \mathbf{B}_{11}(u) \, du \tag{9.41}$$

But

$$\int_{t_i}^{t_{i+1}} \mathbf{Q}_{11}(s, u) \mathbf{B}_{11}(u) \, du = \left[\int_{t_i}^{t_{i+1}} \mathbf{Q}_{11}(s, u) \, du \right] \times \mathbf{B}_{11}^*(t_i, t_{i+1})$$

$$= \mathbf{Q}(s, t_i) \mathbf{E}(t_i, t_{i+1}) \mathbf{B}_{11}^*(t_i, t_{i+1}) \tag{9.42}$$

We thus have

$$\mathbf{M}_{11}(s, t) = \sum_i \mathbf{Q}(s, t_i) \mathbf{E}(t_i, t_{i+1}) \mathbf{B}^*(t_i, t_{i+1}) \tag{9.43}$$

A Remark About the Last Age Interval

When the time axis corresponds to age, the last age interval, say (a_n, ω), invites a remark. It is easily seen that $Q_{11}(a_n, \omega) = 0$. This is so because, by age ω, all individuals reach an absorbing state. Formulas for E_{11}, E_{12} are affected by this condition. Thus,

$$E_{11}(a_n, \omega) = -[R_{11}^*(a_n, \omega)]^{-1} \tag{9.44}$$

Multistate Stationary Populations

Recall the row vector $P(t)$ introduced earlier, with its ith element $P_i(t)$ representing the unconditional probability that a randomly selected individual is in state i at time t. Let us define, analogous to the $_nL_x$ function of the ordinary life table,

$$_hL_t = \int_t^{t+h} P(u)\, du \tag{9.45}$$

which using the relationship that $P(t)$ postmultiplied with $Q(t, t + u)$ gives $P(t + u)$, can be written in the following form:

$$_hL_t = \int_0^h P(t)Q(t, t + u)\, du = P(t)\int_0^h Q(t, t + u)\, du \tag{9.46}$$

Confining attention to the transient states, we have thus

$$_hL_{t(11)} = P(t)\int_0^h Q_{11}(t, t + u)\, du \tag{9.47}$$

Inserting the expression for Q_{11} in (9.47) from (9.29), we obtain on integration

$$_hL_{t(11)} = P(t)[Q_{11}(t, t + h) - I][R_{11}^*(t, t + h)]^{-1} \tag{9.48}$$

We may also use the approximation

$$_hL_t = (h/2)[P(t) + P(t + h)] \tag{9.49}$$

The vector $_hL_t$ gives the stationary population as of date t, classified by state occupied. This interpretation permits defining a matrix whose elements are analogous to the survival ratios one usually computes from the ordinary life table. Note that

$$_hL_{t+h} = \int P(t + h + u)\, du = \int P(t + h)Q(t + h, t + h + u)\, du$$

$$= \left[\int P(t + u)\, du\right]Q^* \tag{9.50}$$

where all integrations are from 0 to h, and $\mathbf{Q}^*$ represents $\mathbf{Q}(t + h, t + h + u)$ evaluated at some u between 0 and h, inclusive. We may write (9.50) in the form

$$_h\mathbf{L}_{t+h} = {_h}\mathbf{L}_t\mathbf{Q}^* \qquad (9.51)$$

which shows that $\mathbf{Q}^*$ is analogous to the matrix of survival ratios ordinarily computed from the life (mortality) table.

2 Applications

We now illustrate the application of multistate life table techniques in the analyses of nuptiality, population redistribution, and labor force participation.

First, a few general remarks about the usual type of data one analyzes are in order. As in the case of the ordinary life table, we may construct a multistate life table using current data for different cohorts or follow-up data for the same cohort. In the former case, the occurrence–exposure rates of transition "observed" for successive age (duration) groups during a particular period (e.g., a calendar year) are assumed to be equal to the corresponding life table rates. The resulting life table portrays the experiences of a hypothetical cohort that goes through what different cohorts go through at different ages (durations) during a particular calendar period.

Follow-up data fall into a number of different types. For example, event history data show the exact times of successive transitions for each individual in a cohort; event count data give the number of transitions of different types occurring in fixed time intervals; and some panel data give only snapshots at fixed successive time points (e.g., place of residence at successive dates of interview.) Of these different types of data, the first one is richer than the rest, because one can compute from such data transition probabilities specific for intervals of any length. When this richness is absent, some transitions may escape notice. Thus, for example, place of residence recorded at 10-year intervals fails to reveal, for those who are residing at the same place at two successive interviews, the history of moves, if any, in the interim. Obviously, this loss of information becomes progressively more serious when the time intervals between points of observations are increasingly separated.

Data obtained through retrospective questions in interview surveys deserve a special comment. The usual practice of collecting retrospective data allows for selectivity, which may lead to biased inferences regarding relationships of interest. This may be the case, for example, when nuptiality data are collected only from ever-married women or when health care data are obtained only for those who are alive at the time of interview. [Selectivity bias has been discussed by, among others, Hoem (1969), Heckman (1979), and Sheps and Menken (1973).]

Nuptiality Analysis

Espenshade (1983), Nour and Suchindran (1984a, 1985), Willekens *et al.* (1982), and others have examined the use of multistate life table techniques in nuptiality analysis. In this section we illustrate the calculations involved and interpret some of the functions of the nuptiality table using data for the female population of the state of North Carolina in 1970.

For this purpose, we distinguish the following states: "never married," "married," "widowed," "divorced," and "dead." Of these, the first four are transient states, and the last one is an absorbing state. From published vital statistics for 1970, counts of marriages, divorces, and deaths by age and marital status were obtained. The number of transitions to "widowhood" was obtained from the number of deaths among married men classified by age of wife. The mid-year population by marital status in each age group was obtained from the 1970 population census. From these data, age-specific rates of transitions from one marital state to another and from each marital state to death were computed using formula (9.25). In applying the formula, the mid-year population was inserted for N_i for evaluating the integral on the right-hand side of (9.25). The rates thus computed were equated to the corresponding life table rates. Table 9.1a shows the rates for the age group 20 to 25. These are the entries of the $\mathbf{R}^*$ matrix pertaining to the age group. The diagonal entries of the matrix were computed from the nondiagonal entries using the condition that the sum of the entries over destination states must be zero.

Notice that the first four rows of the first four columns of $\mathbf{R}^*$ constitute the $\mathbf{R}_{11}^*$ submatrix. The rates pertaining to transition from transient states to the state of death constitute the $\mathbf{R}_{12}^*$ submatrix. Transitions that are logically impossible (e.g., from "married" state to "never married" state, are given zero values.

The matrix $\mathbf{Q}$ of transition probabilities was computed from $\mathbf{R}^*$ using formula (9.29) and the condition that the probabilities of transitions from any given state to that state or other states must sum to unity. In using (9.29), exp($\mathbf{A}$) was approximated by an adequate number of terms of the series

$$\mathbf{I} + \mathbf{A} + (1/2!)\mathbf{A}^2 + \cdots$$

Table 9.1b shows the computed values of the $\mathbf{Q}$ matrix for the age group 20 to 25. The probability that a never-married woman of age 20 will remain never married until age 25 is 0.2678; that of her getting married in the age interval is 0.7215; and so on. To compute the transition probabilities for a wider age interval, formulas such as the following were used:

$$\mathbf{Q}(20, 50) = \mathbf{Q}(20, 25)\mathbf{Q}(25, 30)\mathbf{Q}(30, 35)\mathbf{Q}(35, 40)\mathbf{Q}(40, 45)\mathbf{Q}(45, 50)$$

Table 9.1

Nuptiality Analysis Using Multistate Life Table Methods: North Carolina, 1970

a. Entries of $\mathbf{R}^*(20, 25)$

Origin state	Destination state				
	Never married	Married	Widowed	Divorced	Dead
Never married	−0.2634	0.2626	0	0	0.0008
Married	0	−0.0074	0.0008	0.0060	0.0006
Widowed	0	0.0983	−0.0991	0	0.0008
Divorced	0	0.6385	0	−0.6403	0.0018
Dead	0	0	0	0	0

b. Entries of $\mathbf{Q}(20, 25)$

Origin state	Destination state				
	Never married	Married	Widowed	Divorced	Dead
Never married	0.2678	0.7215	0.0015	0.0054	0.0038
Married	0	0.9850	0.0031	0.0088	0.0031
Widowed	0	0.3836	0.6100	0.0027	0.0037
Divorced	0	0.9367	0.0023	0.0562	0.0048
Dead	0	0	0	0	1

c. Entries of $\mathbf{Q}(20, 50)$

Origin state	Destination state				
	Never married	Married	Widowed	Divorced	Dead
Never married	0.0735	0.7490	0.0405	0.0845	0.0525
Married	0	0.8151	0.0449	0.0929	0.0471
Widowed	0	0.6939	0.1809	0.0741	0.0511
Divorced	0	0.8133	0.0445	0.0932	0.0490
Dead	0	0	0	0	1

Table 9.1c shows the entries of $\mathbf{Q}(20, 50)$. These figures are interpreted in a straightforward fashion. Thus, 7.35 percent of 20-year-old never-married females are expected to remain in that category until age 50, whereas 74.9 percent are expected to be married as of age 50, and so on.

To obtain estimates of the expected lengths of stay in various states during a given age interval of length 5 years (e.g., from 20 to 25 years) formula (9.34) was used, with the requirement that the lengths of stay summed over states must equal 5. The figures obtained by applying (9.34) are shown in Table 9.2a. According to these figures, a never-married woman of age 20 can expect to

Table 9.2

Expected Sojourn Times, Nuptiality Analysis: North Carolina, 1970

a. Entries of $E_{11}(20, 25)$

	Destination state			
Origin state	Never married	Married	Widowed	Divorced
Never married	2.7787	2.1962	0.0028	0.0122
Married	0	4.9513	0.0085	0.0326
Widowed	0	1.0407	3.9441	0.0056
Divorced	0	3.4745	0.0049	1.5066

b. Entries of $E_{11}(20, 50)$

	Destination state			
Origin state	Never married	Married	Widowed	Divorced
Never married	5.5809	22.5062	0.3466	1.0440
Married	0	27.9483	0.4124	1.2000
Widowed	0	17.1125	11.5889	0.8272
Divorced	0	26.2933	0.3965	2.8163

c. Entries of $E_{11}(20, \omega)$

	Destination state				
Origin state	Never married	Married	Widowed	Divorced	Total
Never married	7.6417	35.2445	8.5536	5.2162	56.6560
Ever married	0	41.7560	9.3224	5.7485	56.8269
Widowed	0	29.1922	22.8423	4.6606	56.6951
Divorced	0	40.0718	9.2818	7.3637	56.7173

remain unmarried 2.7787 years in the age interval 20 to 25, and spend 2.1962 years of that age interval in the married state. For the same woman the expected lengths of stay, during ages 20 to 25, in the widowed and divorced states are negligible.

Figures for broader age intervals can be computed using (9.38). Thus

$$E(20, 30) = Q(20, 25)E(25, 30) + E(20, 25)$$

$$E(20, 50) = Q(20, 45)E(45, 50) + E(20, 45)$$

and

$$E(20, \omega) = Q(20, \omega - 5)E(\omega - 5, \omega) + E(20, \omega - 5)$$

where ω is the end of the life span. In practice, the last age group is usually specified as an open interval, e.g., 85 and over. For such open intervals, we compute the expected length of stay using the formula

$$\mathbf{E}_{11}(a+) = -[\mathbf{R}_{11}^*(\omega - a, \omega)]^{-1}$$

The entries of $\mathbf{E}_{11}(20, 50)$ and $\mathbf{E}_{11}(20, \omega)$ computed from the present data are shown in Table 9.2. Notice that the row sums of $\mathbf{E}_{11}(20, \omega)$ provide the expected remaining life for those aged 20 in each of the different marital states. The figures show in this respect very little difference between women in different marital states. The figures in the rows of $\mathbf{E}_{11}$ when expressed as a percent of the respective row sums give the percent distribution over the state space of the remaining life. Thus those never married as of age 20 may expect to spend 13.5 percent of the remaining life in the never-married state, 62.2 percent in the married state, 15.1 percent in the widowed state, and 9.2 percent in the divorced state.

The entries of M computed in accordance with (9.43) and (9.40) are shown in Table 9.3 for age groups 20 to 25, 20 to 50, and 20 +. These figures answer

Table 9.3

Entries of $\mathbf{M}_{11}$

Origin state	Never married	Married	Widowed	Divorced
		Destination state		
a. Between ages 20 and 25				
Never married	0	0.7378	0.0018	0.0132
Married	0	0.0217	0.0040	0.0297
Widowed	0	0.3913	0.0008	0.0062
Divorced	0	0.9624	0.0028	0.0208
b. Between ages 20 and 50				
Never married	0	1.0631	0.0570	0.2225
Married	0	0.1854	0.0652	0.2652
Widowed	0	0.9463	0.0480	0.1754
Divorced	0	1.1707	0.0637	0.2548
c. Between ages 20 and end of life span				
Never married	0	1.1428	0.5632	0.3886
Married	0	0.2672	0.6128	0.4452
Widowed	0	1.0435	0.5323	0.3329
Divorced	0	1.2523	0.6102	0.4344

questions such as the following: For a typical group of 1,000 women who are "never-married" as of age 20 how many marriage spells (including remarriages) may be expected to occur during their remaining life time? [ANSWER: 1,143]. How many divorce spells? [ANSWER: 389]. How many husband deaths? [ANSWER: 563].

Another set of figures of interest concerns how the survivors of an initial 20-year-old cohort, never married as of age 20, distribute themselves over the state space at selected successive time points (e.g., 5, 10, etc., years from the start). To compute such distributions we use the formula

$$\mathbf{P}(t + h) = \mathbf{P}(t)\mathbf{Q}(t, t + h)$$

starting with a distribution that is fully concentrated in the never-married state, e.g., a vector $\mathbf{P}(20)$ that has unity as the entry corresponding to never married and zero everywhere else. The corresponding distribution dated five years later is $\mathbf{P}(25) = \mathbf{P}(20)\mathbf{Q}(20, 25)$, and the one another five years later is $\mathbf{P}(30) = \mathbf{P}(25)\mathbf{Q}(25, 30) = \mathbf{P}(20)\mathbf{Q}(20, 25)\mathbf{Q}(25, 30)$, and so on. The figures thus computed for the data on hand are shown in Table 9.4. Notice that the share going to the "married" state declines after age 35. The reason is that after that age those exits due to divorce, widowhood, and death outnumber those who get married for the first time. From the figures shown in Table 9.4, we can compute the multistate stationary population, using Eq. (9.48), if we let $h = 5$ in the formula. This is left as an exercise.

Working-Life Table

The life table method was first applied to the analysis of labor-force participation only a few decades ago (Wolfbein, 1949; U.S. Department of Labor, 1950; New Zealand, Census and Statistics Department, 1955.) By now

Table 9.4

Proportion of Women in Various States at a Given Exact Age: 20-year-Old Never-Married Women, North Carolina, 1970

Age	Never married	Married	Widowed	Divorced	Dead
20	1.0	0	0	0	0
25	0.2678	0.7215	0.0015	0.0054	0.0038
30	0.1302	0.8421	0.0039	0.0157	0.0081
35	0.1007	0.8529	0.0079	0.0259	0.0128
40	0.0876	0.8287	0.0120	0.0509	0.0208
45	0.0823	0.8001	0.0233	0.0638	0.0325
50	0.0735	0.7490	0.0405	0.0845	0.0525

it has become a standard practice in a number of countries (e.g., Great Britain, United States, New Zealand) to compute complete or abridged working-life tables at frequent intervals of time. Such tables are useful in human resource management, as they reveal important implications of changes in age-specific labor-force participation rates. A common method of construction of working-life tables involves distributing the stationary population of the ordinary life table into active and inactive segments using the participation rates "observed" in a calendar period (e.g., year) and then computing a number of measures such as the average remaining years of active life in the labor force for those who survive to a given age and age-specific rates of accession to and separation from the labor force (the latter sometimes by cause, e.g., death or retirement). Such measures help human resource managers to estimate the replacement needs of industry, the lifetime expected earnings of workers, etc.

The conventional method of construction of a working-life table makes a number of assumptions (Willekens, 1980; Shryock and Siegel, 1973). Among these are that (1) the age-pattern of labor-force participation rate is unimodal, (2) prior to the age at which the labor-force participation rate peaks, there is no separation from the labor force due to retirement or due to move to an inactive state for some other reason, and (3) those in the labor force have the same mortality experiences as the general population. The first two assumptions are usually satisfied for males but very often not for females (Hoem and Fong, 1976). [The female pattern is usually bimodal, due to the impact of factors associated with marriage and childbearing on the age at entry into and exit from the labor force.] Also it may be noted that there is evidence that the mortality experiences of those in the labor force are not the same as those of the general population.

The multistate life table approach to the construction of working-life tables avoids making questionable assumptions such as the ones just mentioned. The Markovian assumption is, however, required for this approach.

We here present a simple working-life table in which two transient states (inactive and active) and one absorbing state (dead) constitute the state space. No new principles are involved if the state space is extended so as to accommodate subdivisions of activity by type of work or mortality by cause of death.

The data used in the illustration are from Hoem and Fong (1976) reported in Willekens (1980). The original source of the material is a Danish labor-force panel survey conducted during 1972–1974. For the present illustration, it is assumed that mortality differentials by activity status are absent. (This assumption is not theoretically required but is a practical necessity dictated by lack of sufficient information to do otherwise.)

The rates of mortality and those of accession to and separation from the

Table 9.5

Age-Specific Labor-Force Participation Rates and Mortality Rates, Denmark, Males, 1973–1974[a]

Age	Inactive to active	Active to inactive	Mortality
16	0.33908	0.52569	0.000733
17	0.60535	0.17170	0.000934
18	0.63573	0.13585	0.001202
19	0.55714	0.12199	0.001327
20	0.45769	0.09226	0.001221
21	0.37317	0.08520	0.001016
22	0.34544	0.08060	0.000944
23	0.34197	0.07150	0.000997
24	0.35514	0.05737	0.001014
25	0.39323	0.04128	0.000932
26	0.42352	0.03225	0.000901
27	0.44979	0.02418	0.000980
28	0.47404	0.01946	0.001033
29	0.49226	0.01509	0.001039
30	0.49939	0.01205	0.001066
31	0.49512	0.01212	0.001198
32	0.47830	0.01253	0.001319
33	0.45148	0.01278	0.001284
34	0.41688	0.01074	0.001260
35	0.37930	0.00937	0.001377

[a] Source: Hoem (1977).

labor force for selected ages are shown in Table 9.5. From these figures we can construct the $\mathbf{R}^*$ matrices in the usual fashion. Thus for $\mathbf{R}^*(25, 26)$ we have

	Inactive	Active	Dead
Inactive	−0.394162	0.393230	0.000932
Active	0.041280	−0.042212	0.000932
Dead	0.0	0.0	0.0

Given $\mathbf{R}^*$, we can construct $\mathbf{Q}$ using the formula

$$\mathbf{Q}(t, t + h) = [\mathbf{I} + 0.5\mathbf{R}^*(t, t + h)][\mathbf{I} - 0.5\mathbf{R}^*(t, t + h)]^{-1} \quad (9.52)$$

which is appropriate when data are available for single years of age, as in the present case. Table 9.6 was prepared in this fashion showing the entries of $\mathbf{Q}$.

Table 9.6

Elements of $\mathbf{Q}(x, x + 1)^a$

Age (x)	1 to 1	1 to 2	2 to 1	2 to 2	1 or 2 to 3
16	0.762690	0.236577	0.366775	0.632492	0.000733
17	0.563450	0.435616	0.123557	0.785509	0.000934
18	0.540524	0.458275	0.097929	0.900869	0.001201
19	0.583244	0.415430	0.090961	0.907712	0.001326
20	0.640191	0.358589	0.072283	0.926496	0.001220
21	0.695673	0.303312	0.069250	0.929734	0.001015
22	0.714525	0.284532	0.066389	0.932668	0.000943
23	0.715877	0.283126	0.059197	0.939807	0.000997
24	0.704844	0.294143	0.047516	0.951470	0.001013
25	0.676296	0.322773	0.033884	0.965185	0.000931
26	0.654463	0.344637	0.026243	0.972856	0.000901
27	0.635724	0.363296	0.019530	0.979490	0.000980
28	0.619101	0.379867	0.015594	0.983373	0.001032
29	0.606674	0.392287	0.012025	0.986936	0.001039
30	0.601623	0.397312	0.009587	0.989348	0.001063
31	0.604275	0.394527	0.009658	0.989145	0.001198
32	0.615089	0.383592	0.010049	0.988633	0.001319
33	0.632720	0.365997	0.010360	0.988357	0.001283
34	0.655688	0.343053	0.008838	0.989900	0.001259
35	0.681440	0.317181	0.007835	0.990789	0.001376

^{a} Inactive, active, and dead states are coded 1, 2, and 3, respectively.

Thus the entries for $\mathbf{Q}(25, 26)$ are

	Inactive	Active	Dead
Inactive	0.6763	0.3228	0.0009
Active	0.0339	0.9652	0.0009
Dead	0.0	0.0	0.0

The interpretation of these figures is straightforward. Thus the probability of an inactive person who has just turned 25 remaining inactive until age 26 is 0.6763. Using the matrices $\mathbf{Q}(t, t + 1)$ for successive age intervals, we compute $\mathbf{Q}(t, t + n)$, for integer n, by multiplication:

$$\mathbf{Q}(t, t + n) = \mathbf{Q}(t, t + 1)\mathbf{Q}(t + 1, t + 2) \cdots \mathbf{Q}(t + n - 1, t + n)$$

Table 9.7 contains the elements of submatrices of cumulative probabilities of transition between the transient states (inactive and active). Thus, from the

Table 9.7

Elements of $\mathbf{Q}_{11}(16, x)^a$

Age (x)	1 to 1	1 to 2	2 to 1	2 to 2
17	0.762690	0.236577	0.366775	0.632492
18	0.563446	0.435613	0.123565	0.875525
19	0.347211	0.650652	0.152506	0.845327
20	0.261699	0.734847	0.165851	0.830683
21	0.220656	0.774667	0.166241	0.829124
22	0.207150	0.787166	0.173043	0.821282
23	0.200269	0.793106	0.178155	0.815216
24	0.190324	0.802070	0.175816	0.816602
25	0.172262	0.819124	0.162731	0.828647
26	0.144255	0.846200	0.138122	0.852346
27	0.116611	0.872947	0.112777	0.876782
28	0.091184	0.897420	0.088818	0.899788
29	0.070448	0.917135	0.069018	0.918548
30	0.053770	0.932791	0.052901	0.933625
31	0.041285	0.944214	0.040769	0.944716
32	0.034068	0.950249	0.033751	0.950565
33	0.030507	0.952520	0.030328	0.952688
34	0.029163	0.952587	0.029072	0.952688
35	0.027550	0.952977	0.027468	0.953035

[a] Inactive and active states are coded 1 and 2, respectively.

last row of Table 9.7 we have the following entries for $\mathbf{Q}_{11}$ (16, 35):

	Inactive	Active
Inactive	0.027550	0.952977
Active	0.027468	0.953035

Table 9.8 contains the elements of the submatrices $\mathbf{E}_{11}(t, t + 1)$ computed using formula (9.34). Thus, from the entries for age 25 we have the rows and columns of $\mathbf{E}_{11}(25, 26)$ as shown below:

	Inactive	Active
Inactive	0.838148	0.161386
Active	0.016942	0.982593

Table 9.8

Elements of $\mathbf{E}_{11}(x, x + 1)^a$

Age (x)	e_{11}	e_{12}	e_{21}	e_{22}
17	0.781725	0.217808	0.061779	0.937755
18	0.770262	0.229137	0.048965	0.950435
19	0.791622	0.207715	0.045481	0.953856
20	0.820096	0.179295	0.036142	0.963248
21	0.847836	0.151656	0.034625	0.964867
22	0.857262	0.142266	0.033194	0.966334
23	0.857939	0.141563	0.029598	0.969903
24	0.852422	0.147071	0.023758	0.975735
25	0.838148	0.161386	0.016942	0.982593
26	0.827231	0.172318	0.013122	0.986428
27	0.817862	0.181648	0.009765	0.989745
28	0.809550	0.189934	0.007797	0.991687
29	0.803337	0.196144	0.006013	0.993468
30	0.800811	0.198656	0.004793	0.994674
31	0.802138	0.197264	0.004829	0.994573
32	0.807545	0.191796	0.005024	0.994316
33	0.816360	0.182998	0.005180	0.994178
34	0.827844	0.171527	0.004419	0.994951
35	0.840721	0.158591	0.003918	0.995394

[a] Active and inactive states are coded 1 and 2, respectively.

indicating that an inactive person who has just turned 25 can expect to spend on average 83.8 percent of the next year in the inactive state and 16.1 percent in active state, and, similarly, an active person who has just turned 25 can expect to remain active 98.3 percent of the next year and inactive for 1.7 percent of the year.

The expected lengths of remaining inactive and active lives for persons surviving to a given age, computed using formula (9.44), are shown in Table 9.9. The figures (the elements of $\mathbf{E}_{11}$) for 17 year olds are

	Inactive	Active
Inactive	10.09	41.59
Active	8.71	42.97

indicating that an inactive person who has just turned 17 can expect to spend during his remaining lifetime about 10 years in the inactive state and about 41.6 years in the active state. The corresponding figures for an active person who has just turned 17 are 8.7 and 43 years, respectively.

Table 9.9

Elements of $\mathbf{E}_{11}(x, \omega)^a$

Age (x)	e_{11}	e_{12}	e_{21}	e_{22}
17	10.09	41.59	8.71	42.97
18	9.32	41.41	8.66	42.07
19	8.87	40.91	8.53	41.26
20	8.58	40.27	8.38	40.47
21	8.35	39.56	8.22	39.69
22	8.14	38.82	8.06	38.89
23	7.94	38.06	7.89	38.11
24	7.76	37.29	7.72	37.33
25	7.58	36.51	7.56	36.54
26	7.43	35.71	7.42	35.72
27	7.30	34.87	7.30	34.87
28	7.20	34.01	7.20	34.01
29	7.13	33.12	7.13	33.12
30	7.07	32.22	7.07	32.22
31	7.03	31.31	7.03	31.31
32	7.00	30.38	7.00	30.38
33	6.98	29.45	6.98	29.45
34	6.96	28.52	6.96	28.52
35	6.94	27.58	6.94	27.58

[a] Inactive and active states are coded 1 and 2, respectively.

One can construct the **M** matrix as illustrated below, using the formula $\mathbf{M}_{11}(17, 18) = \mathbf{E}_{11}(17, 18)\mathbf{B}_{11}^*(17, 18)$:

$$\begin{bmatrix} 0.781725 & 0.217808 \\ 0.061729 & 0.937755 \end{bmatrix}\begin{bmatrix} 0 & 0.33908 \\ 0.52569 & 0 \end{bmatrix} = \begin{bmatrix} 0.265067 & 0.114499 \\ 0.020948 & 0.492968 \end{bmatrix}$$

According to these figures, among 1000 inactive persons who have just turned 17, we can expect during the age interval 17 to 18, 1145 spells of active life and 265 spells of inactive life, and, similarly, among 1,000 active persons who have just turned 17, we can expect, during the same period, 21 inactive spells and 493 active spells. There are of course other summary measures one can compute from the figures given. These are left as exercises.

Multi-regional Life Tables

Multistate life table techniques can be used to describe population redistribution over a given system of territorial units (e.g., states or regions within a country). Using such tables one can study migration patterns as well as make population projections for subnational areas (Philipov and Rogers, 1982).

Here we illustrate the construction of multiregional life tables using age-specific migration and death rates for U.S. females for 1968, shown in Table 9.10. These are taken from Philipov and Rogers (1982). Only two regions are distinguished: north and south. The state space thus consists of "north," "south," and "dead." Table 9.10 permits the construction of R^* matrices as illustrated below for the age group 15 to 20:

	South	North	Dead
South	-0.019752	0.018979	0.000773
North	0.008783	-0.00944	0.000657
Dead	0	0	0

Table 9.11 shows the transition probabilities computed from the R^* matrices using Eq. (9.52) introduced earlier in connection with the construction of a

Table 9.10

Age-Specific Migration and Death Rates for Females, 1968[a,b]

Age (x)	S to N	S to D	N to S	N to D
0	0.011977	0.005332	0.005620	0.004925
5	0.008685	0.000425	0.004403	0.000364
10	0.008875	0.000348	0.004522	0.000281
15	0.018835	0.000641	0.008963	0.000563
20	0.018979	0.000773	0.008783	0.000657
25	0.012627	0.000949	0.006813	0.000779
30	0.008751	0.001397	0.004900	0.001109
35	0.006772	0.002255	0.004000	0.001824
40	0.005147	0.003315	0.003256	0.002817
45	0.003497	0.004753	0.002536	0.002817
50	0.002978	0.006564	0.002955	0.006159
55	0.002605	0.009513	0.004112	0.009054
60	0.003245	0.013319	0.005375	0.013352
65	0.003170	0.020997	0.003908	0.021110
70	0.003754	0.031955	0.002760	0.034095
75	0.003227	0.053139	0.002234	0.056736
80	0.001335	0.085379	0.000891	0.089561
85	0.000637	0.149437	0.000435	0.157495

[a] N = north, S = south, D = dead.
[b] Source: Philipov and Rogers (1982)

Table 9.11

Elements of $\mathbf{Q}(x, x + 5)^a$

Age (x)	S to N	S to S	S to D	N to S	N to N	N to D
0	0.055950	0.917796	0.026254	0.026252	0.949396	0.024352
5	0.041968	0.955917	0.002114	0.021276	0.967904	0.001820
10	0.042870	0.955400	0.001730	0.021842	0.976753	0.001405
15	0.087799	0.909020	0.003182	0.041781	0.955401	0.002817
20	0.088430	0.907739	0.003830	0.040924	0.955783	0.003293
25	0.059957	0.935336	0.004707	0.032351	0.963750	0.003899
30	0.042050	0.951020	0.006930	0.023546	0.970906	0.005548
35	0.032643	0.956178	0.011179	0.019283	0.971616	0.009101
40	0.024827	0.958764	0.016409	0.015707	0.970285	0.014008
45	0.016849	0.959691	0.023460	0.012222	0.967205	0.020573
50	0.014220	0.953504	0.032276	0.014110	0.955548	0.030341
55	0.012239	0.941315	0.046446	0.019321	0.936389	0.044290
60	0.014885	0.920663	0.064451	0.024653	0.910747	0.064600
65	0.014068	0.886180	0.099752	0.017345	0.882401	0.100253
70	0.015780	0.836188	0.148032	0.011600	0.831374	0.157026
75	0.012326	0.753037	0.234636	0.008532	0.743096	0.248372
80	0.004474	0.643685	0.351842	0.002987	0.631134	0.365879
85	0	0	1.0	0	0	1.0

[a] N = north, S = south, D = dead.

working-life table. Thus for the age group 20 to 25 we have for the entries of $\mathbf{Q}(20, 25)$

	South	North	Dead
South	0.907739	0.088430	0.003830
North	0.040924	0.955783	0.003293
Dead	0	0	1

indicating, for example, that for a woman who has just turned 20 and resides in the south, the probability of residing in the north on her 25th birthday is 0.08843, and of dying before age 25 is 0.00383.

Transition probabilities covering longer periods are presented in Table 9.12. These are the elements of $\mathbf{Q}_{11}(0, x)$. The elements of $\mathbf{Q}_{11}(0, 20)$ are

	South	North
South	0.77021	0.19681
North	0.09575	0.87397

Table 9.12
Elements of $\mathbf{Q}_{11}(0, x)^a$

Age (x)	S to S	S to N	N to N	N to S
5	0.91780	0.05595	0.94940	0.02625
10	0.87853	0.09318	0.92857	0.04529
15	0.84138	0.12867	0.90893	0.06355
20	0.77021	0.19681	0.87397	0.09575
25	0.70720	0.25621	0.84379	0.12268
30	0.66976	0.28933	0.82056	0.14205
35	0.64377	0.30907	0.80266	0.15441
40	0.62152	0.32132	0.78492	0.16312
45	0.60093	0.32720	0.76564	0.16872
50	0.58071	0.32659	0.74338	0.17128
55	0.55832	0.32033	0.71277	0.17380
60	0.53174	0.30679	0.66955	0.17738
65	0.49712	0.28732	0.61243	0.17981
70	0.44552	0.26053	0.54294	0.16997
75	0.37556	0.22363	0.45407	0.14842
80	0.28472	0.17080	0.33925	0.11564
85	0.18378	0.10907	0.21463	0.07545

[a] N = north, S = south.

indicating that a new-born girl in the south has a probability of 0.19681 of residing in the north on her 20th birthday, while her counterpart born in the North has a probability of 0.09575 residing in the south on her 20th birthday. [Note that the probability of death can be calculated by subtracting from unity the sum of the probabilities of residing in the respective regions. Thus, for a new-born southern girl, the probability of dying before 20th birthday is 0.03298 ($= 1 - 0.19681 - 0.70720$)].

The elements of $\mathbf{E}_{11}$ matrices can be computed in the usual fashion. Thus for the age interval 20 to 25 we have for the elements of $\mathbf{E}_{11}$

	South	North
South	4.76430	0.22445
North	0.10681	4.88714

indicating that a woman in the south who has just turned 20 can expect to spend 4.76 years during the next five years in the south and 0.22 years in the north, while her counterpart residing in the north on her 20th birthday can expect to spend 0.10 years in the south and 4.88 years in the north.

The cumulative lengths of stay in each transient state are shown in Table 13. Thus for the entire life span 0 to ω the entries of $\mathbf{E}_{11}$ are

	South	North
South	53.81	20.34
North	11.29	63.16

according to which a new-born southern girl may expect to spend during her lifetime 53.8 years in the south and 20.34 years in the north, while her counterpart born in the north can expect to spend 11.29 years in the south and 63.16 years in the north. These can of course be expressed as percentages of the life expectancy at birth. Thus a new-born southern girl can expect to spend 73 percent of her lifetime in the south, and so on.

The computation of the $\mathbf{M}$ matrices is straightforward. Thus for the age interval 20 to 25 we have the $\mathbf{M}_{11}$ matrix given by

$$\begin{bmatrix} 4.76430 & 0.22445 \\ 0.10681 & 4.88714 \end{bmatrix} \begin{bmatrix} 0 & 0.018979 \\ 0.008783 & 0 \end{bmatrix} = \begin{bmatrix} 0.04184 & 0.00197 \\ 0.00197 & 0.09275 \end{bmatrix}$$

indicating that among, say, 1,000 females who have just turned 20 in the south, we can expect on the average 20 moves to (or different spells of stay in) the north during the next five years, while among their counterparts (numbering 1,000) we can expect on average an equal number of moves to the south.

Problems and Complements

1. *Working life tables*: Early works on the application of multistate life table approach to the analysis of labor-force participation include those of Hoem (1977), and Willekens (1980). Since 1982 the U.S. Bureau of Labor Statistics (Department of Labor) has been constructing working-life tables using the multistate approach [U.S. Department of Labor, Bureau of Labor Statistics (1982a,b); Smith and Horvath (1984).]

2. *Nuptiality*: More than ten years ago, Schoen (1975) presented one of the earlier multistate nuptiality tables. Krishnamoorthy (1979a), Espenshade (1983, 1985), Espenshade and Braun (1982), Hofferth (1985a,b) Schoen and Baj (1983), and Schoen et al. (1985) are among those who have applied multistate life table methodology to the analysis of nuptiality in the United States. Krishnamoorthy (1982) has applied the methodology to Australian data and Koesoebjono (1979) to The Netherlands. A cross-national comparative analysis involving the United States, Belgium, England and Wales, Sweden, and Switzerland can be found in Schoen and Baj (1983).

3. *Other applications*: An application of the multistate life table approach to the analysis of voting status in the U.S. presidential elections can be found in Land and Hough (1985). Haberman (1983) has discussed possible applications of multistate life table techniques in morbidity analysis. In a recent collection of articles edited by Bongaarts et al. (1985), the application of the multistate

life table approach to the study of family demography has been discussed by a number of scholars. Rogers (1966, 1975) and following him many others (e.g., Rees and Wilson, 1973, 1977; Willekens and Drewe, 1984; and Ledent and Rees, 1986) have applied the methodology of multistate life tables to migration analysis.

4. Schoen *et al.* (1985) presents a number of summary measures pertaining to multistate life tables (with specific reference to nuptiality.) To give one example, in a multistate stationary population, the number of transitions from state i to state j in a given time interval $(x, x + h)$ is given by

$$d_{ij}(x, x + h) = P_i \times q_{ij}(x, x + h)$$

If we assume that transitions take place at the middle of the interval $(x, x + h)$, the mean duration of stay at which transition from state i to state j takes place is easily computed for the entire lifespan for those who do make the transition.

5. Follow-up data usually permit direct computation of the **Q** matrix. Let $D_{ij}(x, x + h)$ be the (observed) numbers of transitions from state i to state j, over the interval $(x, x + t)$, $N_i(x)$ the number of individuals in state i at age x, and $W_i(x)$ the number withdrawn from state i during $(x, x + t)$. Then

$$q_{ij}(x, x + t) = D_{ij}(x, x + t)/(N_i(x) - \tfrac{1}{2} W_i(x))$$

When the state space contains only transient states, the **R** matrix is obtained from the relation

$$\mathbf{R} = -(1/t)[(\mathbf{I} - \mathbf{Q}) + (\mathbf{I} - \mathbf{Q})^2/2 + \cdots]$$

where **I** is the identity matrix. For an application see (Griffith *et al.*, 1985). From the marital history data provided in the 1980 June supplement of the *U.S. Current Population Survey*, the following entries of **Q**(20, 25) were computed (Griffith *et al.*, 1985):

	Never married	Married	Separated	Divorced	Widowed
Never married	0.3590	0.6055	0.0119	0.0194	0.0025
Married	0	0.9117	0.0230	0.0586	0.0067
Separated	0	0.5856	0.0358	0.3759	0.0029
Divorced	0	0.7120	0.0178	0.2673	0.0039
Widowed	0	0.6043	0.0167	0.0306	0.3488

Compute the corresponding **R** matrix and the **E** and **M** matrices. The entries of the **E** matrix are

	Never married	Married	Separated	Divorced	Widowed
Never married	3.1337	1.7938	0.0273	0.0342	0.0044
Married	0	4.7186	0.1041	0.1583	0.0191
Separated	0	1.5222	1.2960	2.1776	0.0047
Divorced	0	2.3387	0.0563	2.6017	0.0073
Widowed	0	1.8480	0.0553	0.0540	3.0427

6. Using data from the 1972–1973 January *U.S. Current Population Survey*, and the fourth edition of the *Dictionary of Occupational Titles*, Hayward *et al.* (1985) applied multistate life table

techniques to examine how the nature of work influences the labor force behavior of older men and women. Obviously, in such exercises the labor force states are expanded to incorporate the type of work (occupation).

Bibliographic Notes

Multistate demography, according to some experts, is in the process of becoming a subdiscipline by itself (Willekens, 1985b). There is every indication that the comparative static approach based on participation rate, as in the traditional labor force analysis, will be replaced by a dynamic approach based on the notion of transition (mobility). Developments in the analysis of transition matrices together with the advances in survival analysis have already transformed the landscape of multistate demography into an exciting field. Some familiarity with matrix algebra is essential to follow the literature. An introduction is available in Namboodiri (1984). Some familiarity with stochastic processes, particularly Markov processes, is also essential to understand the logic of multistate demography. A good reference in this connection is Bhat (1984). Keyfitz (1980) has provided an assessment of where multistate demography is headed. Willekens (1985b) traces the beginning of multistate demography to Rogers' works on multiregional demography. Nour and Suchindran (1985), among others, provide a quick review of the theory of multistate life tables. The collection of papers edited by Land and Rogers (1982) give an indication of the growth potential of the subfield. References cited in the third paragraph of bibliographic note 1 are all relevant here.

Chapter 10 | Survival Distributions

In this chapter a number of concepts, definitions, models, and statistical methods pertaining to the analysis of survival time are briefly reviewed. Survival time, it may be recalled, is a loosely defined term applied to such things as the time to failure of a physical component, the time to death of a biological entity (e.g., a human individual or an animal cell), the time taken by children to master a skill, the duration of residence at an address before the first move, the interval between divorce and remarriage, or the time to the demise or dismantling of a political arrangement or an organization. The term "survival analysis" is applied to a variety of techniques treating survival time as a positive-valued random variable. The origin of survival analysis dates back to the early works on mortality tables. But the more modern literature on the subject started to develop in the 1930s with applications to industrial engineering and to reliability studies of military equipment in connection with World War II. More recently, the field of application has been widened to include medical and demographic studies and survey analysis.

1 Basic Concepts of Survival Time Distributions

When survival time is regarded as a random variable, it is customary to denote it by a capital letter, e.g., $T > 0$, and use the corresponding lower-case letter to denote a typical value taken by the random variable. But the custom is not always followed. For the moment, we use T for the random variable of interest. Our present objective is to describe the various functions of T that are used in survival analysis. Unless otherwise stated all functions are defined for positive values of the random variable.

First, we introduce what is known as the *survival function* (or *survivor function*) $S(t)$, to denote the probability that $T > t$:

$$S(t) = \Pr(T > t) \tag{10.1}$$

Here note that T is positive, and $S(t)$ is a nonincreasing function with $S(0) = 1$ and $S(\infty) = 0$. The complement of $S(t)$, i.e., $1 - S(t)$, is often denoted by $F(t)$, and is known by the name *cumulative distribution function* or simply the *distribution function*.

Another function of interest is the *probability density function* (pdf) denoted by $f(t)$, which is such that

$$\Pr(t \le T \le t + \Delta t) = \int_t^{t+\Delta t} f(u) \, du \tag{10.2}$$

Clearly

$$S(t) = \int_t^{\infty} f(u) \, du \tag{10.3}$$

and

$$F(t) = \int_0^t f(u) \, du \tag{10.4}$$

Also of interest is the *hazard (rate) function* $h(t)$, defined as

$$h(t) = \lim_{\Delta t \to 0} \left[\frac{\Pr(t < T < t + \Delta t \,|\, T > t)}{\Delta t} \right]$$

$$= \lim_{\Delta t \to 0} \left[\int_t^{t+\Delta t} f(u) \, du \Big/ \int_t^{\infty} f(u) \, du \right]$$

$$= f(t)/S(t) \tag{10.5}$$

[Some authors use $\lambda(t)$ instead of $h(t)$. In mortality analysis, $\mu(t)$, or more frequently μ_t, is used instead.]

The functions $f(t)$, $F(t)$, $S(t)$, and $h(t)$ give mathematically equivalent descriptions of the random variable T. It is easy to recognize from (10.3) that

$$S'(t) = -f(t) \tag{10.6}$$

and hence from (10.5) that

$$h(t) = -S'(t)/S(t) \tag{10.7}$$

which on integration gives

$$-\int_0^t h(u) \, du = \ln S(t) \Big|_0^t = \ln S(t) \tag{10.8}$$

the last step following from the property that $S(0) = 1$. We thus have

$$S(t) = \exp\left[-\int_0^t h(u)\, du\right] \tag{10.9}$$

and

$$f(t) = h(t)S(t) = h(t) \exp\left[-\int_0^t h(u)\, du\right] \tag{10.10}$$

For some purposes,

$$H(t) = \int_0^t h(u)\, du = -\ln S(t) \tag{10.11}$$

known as the *cumulative hazard function* (note the absence of the term "rate"), is used. It is easy to recognize that

$$S(t) = \exp[-H(t)] \tag{10.12}$$

2 Some Survival Distributions

In this section, we mention several well-known survival distributions. For extensive treatment of these the reader is referred to Lawless (1982) in particular.

Exponential Distribution

The pdf of an exponential distribution

$$f(t) = \alpha \exp(-\alpha t), \qquad \alpha > 0 \tag{10.13}$$

When $\alpha = 1$, we have $f(t) = \exp(-t)$, which is known as the *standard exponential distribution*. Figure 10.1 shows the shape of the standard exponential.

In the general case,

$$S(t) = \int_t^\infty f(u)\, du = \exp(-\alpha t) \tag{10.14}$$

and

$$h(t) = f(t)/S(t) = \alpha \tag{10.15}$$

Thus the exponential distribution with pdf (10.13) is characterized by a constant hazard rate α. This distribution has mean $1/\alpha$ and variance $(1/\alpha)^2$.

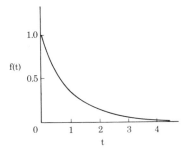

Fig. 10.1. The standard exponential probability density function.

The *coefficient of variation*, i.e., the ratio of the mean to the standard deviation, is unity. This serves as a reference standard for judging relative variance of other distributions.

Historically, the exponential distribution was the first lifetime model for which statistical methods were extensively developed (see Galambos and Kotz, 1978; Johnson and Kotz, 1970). Sukhatme (1937) and Epstein and Sobel (1953) are among those who were responsible for popularizing this model in applied work.

In real situations, the analyst often segments the lifetime into a set of small intervals such that in each interval one may assume that the hazard rate is constant. This strategy is frequently used in applied work, as will become clear later on.

The Extreme-Value Distribution

In modeling such phenomena as the occurrence of floods and droughts and the failure of mechanical systems consisting of a number of components, the theory of extreme values and of extreme-value distributions have important applications. Thus, if we make the reasonable assumption that in a chain the first link that breaks is the one with the least strength, we may represent the strength of the chain as the minimum of the random variables representing the strengths of the constituent links. On the other hand, if twin-engined airplanes fail only if both engines fail, then the failure of such airplanes corresponds to the failure of the strongest engine. The pdf and the survival function of the (least) extreme-value distribution are

$$f(t) = \frac{1}{\beta} \exp\left[\frac{t-u}{\beta} - \exp\frac{t-u}{\beta}\right] \qquad (10.16)$$

$$S(t) = \exp\left[-\exp\frac{t-u}{\beta}\right] \qquad (10.17)$$

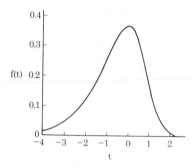

Fig. 10.2. The standard extreme-value probability density function.

respectively, where t and u may lie anywhere on the real number line and $\beta > 0$. We will find later on that the special case of the distribution just described, corresponding to $u = 0$ and $\beta = 1$ (known as the *standard extreme value distribution*), is useful in specifying the distribution of the error terms in regression models for lifetimes. The mean of the standard extreme value distribution is Euler's constant $(=0.5772\ldots)$ and the variance is $\pi^2/6$. Figure 10.2 shows the shape of the standard extreme-value distribution.

The Gompertz and Makeham–Gompertz Distributions

Notice that the extreme-value distribution described above is defined for the entire real number line. To represent lifetimes, which are always positive valued, one uses the extreme-value distribution truncated from below at $t = 0$. The survival function for the distribution thus truncated is

$$S(t) = \exp[(b/a)(1 - e^{ax})] \tag{10.18}$$

where $a > 0$ and $b > 0$. The corresponding hazard function is

$$h(t) = b \exp(ax) \tag{10.19}$$

This is known as the *Gompertz distribution*, more commonly written in the following form:

$$h(t) = bc^x \tag{10.20}$$

This is obtained from (10.19) by setting $c = \exp(a)$. This distribution is widely used in actuarial work (to represent mortality patterns; see in this connection, Wilkin, 1981, for a report on a study that indicates that Gompertz' law does not accurately represent the observed overall pattern of mortality above age

90). If a constant is added to the hazard function (10.20), to allow for accidental deaths in addition to deaths from natural causes one gets

$$h(t) = a + bc^x \tag{10.21}$$

which is known as the *Makeham–Gompertz distribution.*

The Log-Normal Distribution

The lifetime T is log-normally distributed if the natural logarithm of T is normally distributed. If $Y = \ln T$ is normally distributed with mean μ and variance σ^2, then the pdf of Y is

$$(1/\sqrt{2\pi\sigma^2}) \exp[-(y - \mu)^2/2\sigma^2] \tag{10.22}$$

and the pdf of $T = \exp(Y)$ is given by (see Fig. 10.3)

$$f(t) = (1/\sqrt{2\pi\sigma^2 t^2}) \exp[-(\ln t - \mu)^2/2\sigma^2] \tag{10.23}$$

The log-normal distribution has been widely used for modeling lifetime [e.g., in the study of times to the appearance of lung cancer among cigarette smokers (see Whittemore and Altschuler, 1976; also see Gaddum, 1945a,b; Boag, 1949; Aitchison and Brown, 1957; and Feinleib, 1960).

The survival function of the log-normal distribution is

$$S(t) = 1 - \Phi[(\ln t - \mu)/\sigma] \tag{10.24}$$

where $\Phi(x)$ stands for the area to the left of x under the standard normal. The corresponding hazard function is obtained from the relationship $h(t) = f(t)/S(t)$.

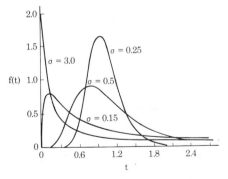

Fig. 10.3. Log-normal probability density functions with $\mu = 0$ and $\sigma = 0.25, 0.5, 1.5,$ and 3.0.

One major drawback of the log-normal distribution is that its hazard function is not monotonic—it starts with the value 0 at $t = 0$, increases to a maximum, and then decreases to 0. Since the hazard function is decreasing for large values of t, the distribution seems implausible in most common situations as a model for lifetime. Nevertheless, as mentioned above, this distribution has been found suitable for representing lifetime data in many practical situations (especially those in which large values of lifetime are not of interest.) Cox and Oakes (1984) point out that unless substantial amount of data are on hand, it would be impossible to discriminate the exponential from the log-normal with $\sigma = 0.8$.

The Weibull Distribution

This is one of the most widely used models in survival analysis. Weibull (1951) discusses its wide applicability. Lieblein and Zelen (1956) apply it to the study of the fatigue life of deep-groove ball bearings; Pike (1966), Peto et al. (1972), Peto and Lee (1973), and Williams (1978) among others use it in the analysis of data from carcinogenesis experiments.] The pdf of the Weibull distribution is

$$f(t) = \alpha\beta(\alpha t)^{\beta-1} \exp[-(\alpha t)^{\beta}] \tag{10.25}$$

and the corresponding survival and hazard functions are

$$S(t) = \exp[-(\alpha t)^{\beta}] \tag{10.26}$$

and

$$h(t) = \alpha\beta(\alpha t)^{\beta-1} \tag{10.27}$$

respectively. The Weibull hazard function is increasing if $\beta > 1$, decreasing if $\beta < 1$; and constant if $\beta = 1$ (see Fig. 10.4.) In the last case, the Weibull is the same as the exponential.

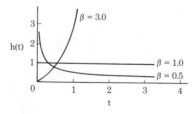

Fig. 10.4. Weibull hazard functions with shape parameter $= 0.5, 1.0, 3.0$, location parameter in each case being 0.

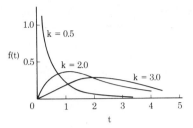

Fig. 10.5. One-parameter gamma distribution with probability density function $[\Gamma(k)]^{-1}t^{k-1}e^{-t}$.

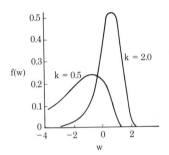

Fig. 10.6. Log-gamma distribution with probability density function $[\Gamma(k)]^{-1}\exp(k\omega - e^{-\omega})$.

Other Distributions

Among the other distributions commonly discussed in the literature on survival analysis are the gamma (see Fig. 10.5), the log-gamma (see Fig. 10.6), the generalized gamma, the inverse Gaussian, the logistic, and the log-logistic distributions. The reader is referred to Lawless (1982), Cox and Oakes (1984), and Johnson and Kotz (1970) for discussions of these.

3 Maximum Likelihood Estimation and Likelihood Ratio Tests

In this section the reader's attention is drawn to (1) construction of likelihood function, (2) the maximum likelihood method of estimation, and (3) the use of likelihood ratio tests in assessing "goodness of fit."

Construction of Likelihood Function

A statistical model for a random variable specifies how the latter is generated. The observed data are viewed as realizations of the process thus specified. Given a statistical model and a set of data, the likelihood function is proportional to the joint probability (density) for the observed data regarded as a function of the parameters of the statistical model. To give a simple example, suppose ten patients selected at random from a population of patients are treated with an experimental drug. Suppose four of the ten are cured. In this case we have ten random variables, one for each patient, taking the value, say, 1, if the patient was cured and 0 otherwise. The realized values of these ten random variables in this particular case consist of four ones and six zeros. If we assume that the sampling is binomial, the joint probability for the outcome of the experiment is proportional to

$$\pi^4(1 - \pi)^6 \tag{10.28}$$

where π is the probability of getting a 1 in a single trial. The expression (10.28) as a function of the parameter π is the likelihood function on the data.

To give another example, suppose starting from time 0, n individuals are observed until all fail. Let t_i be the exact failure time of the ith individual, $i = 1, 2, \ldots, n$. Suppose the failure model has $f(t, \alpha, \beta)$ for its pdf, where α and β are unknown parameters. Then if the failure times are mutually independent, the likelihood function on the data for the two parameters α and β is given by

$$\text{lik} = f(t_1, \alpha, \beta) \cdot f(t_2, \alpha, \beta) \cdots f(t_n, \alpha, \beta) \tag{10.29}$$

To give a third example, suppose starting from time 0, n individuals are observed, but by the end of the observation period only r individuals had failed. Then for the remaining $(n - r)$ individuals we know only that all of them survived (not failed) until the end of the observation period. Let the subjects who failed be indexed $1, 2, \ldots, r$, and let their exact failure times be $t_1, t_2, \ldots, t_r$, respectively. Let t_c be the time of termination of the observation period. If the pdf of the failure model is $f(t, \alpha, \beta)$ and the survival function is $S(t, \alpha, \beta)$, the likelihood on the data for the parameters α and β is proportional to the product

$$f(t_1, \alpha, \beta) \cdots f(t_r, \alpha, \beta)[S(t_c, \alpha, \beta)]^{n-r} \tag{10.30}$$

The Maximum Likelihood Method of Estimation

If there is a set of values for the parameters for which the likelihood is maximum, then the values in the set constitute the *maximum likelihood estimates* of the parameters involved. Different data may yield different maximum likelihood estimates. When sampling variability is the focus the term *maximum likelihood estimators* (MLEs) is used instead of maximum likelihood estimates.

In practice, one usually maximizes the logarithm of the likelihood rather than the likelihood itself. This is justifiable because the log likelihood is a monotonic (increasing) function of the likelihood, and consequently the values of the parameters that maximize one also maximize the other.

Example. Suppose in a life testing experiment, 13 items were involved, but the experiment was terminated when the tenth item failed. Suppose the failure times observed were

$$0.2, \quad 0.5, \quad 0.9, \quad 1.0, \quad 1.3, \quad 1.4, \quad 1.5, \quad 1.8, \quad 2.5, \quad 3.0$$

For the remaining three items, we know only that their failure times exceeded 3.0. Let us assume that the Weibull distribution applies here. Recall that the Weibull pdf is

$$\alpha\beta(\alpha t)^{\beta-1} \exp[-(\alpha t)^{\beta}] \tag{10.31}$$

It is more convenient to work with the pdf of $X = \ln T$. If T has the Weibull pdf (10.31), then $X = \ln T$ has the pdf

$$f(x) = \exp\{(x - u)/b\} \exp\{-\exp[(x - u)/b]\} \tag{10.32}$$

where $\beta = 1/b$ and $\alpha = \exp(-u)$. The survival function corresponding to (10.32) is

$$S(x) = \exp\{-\exp[(x - u)/b]\} \tag{10.33}$$

Denoting the transformed observations by $x_1 = \ln 0.2$, $x_2 = \ln 0.5, \ldots, x_{10} = \ln 3.0$, the likelihood on the data for the parameters u and b is

$$\text{lik} = f(x_1) \cdots f(x_{10})[S(x_{10})]^3 \tag{10.34}$$

The log likelihood is

$$L = -10 \ln b + \sum W_i - \sum{}^* \exp(W_i) \tag{10.35}$$

where

$$W_i = (x_i - u)/b$$

and

$$\sum{}^* y_i = y_1 + \cdots + y_{10} + 3y_{10}$$

Taking partial derivatives of L with respect to the parameters u and b gives

$$\frac{\partial L}{\partial u} = -\frac{10}{b} + \frac{1}{b}\sum{}^* \exp(W_i) \tag{10.36}$$

and

$$\frac{\partial L}{\partial b} = \frac{1}{b}\left[-10 - \sum W_i + \sum{}^* W_i \exp(W_i)\right] \tag{10.37}$$

The maximum likelihood estimates of u and b are given by the solutions to the equations $\partial L/\partial u = \partial L/\partial b = 0$. Setting (10.36) equal to zero gives

$$\exp(u) = \left[\frac{1}{10} \sum{}^* \exp \frac{x_i}{b} \right]^b \tag{10.38}$$

and setting (10.37) equal to zero gives on simplification and substitution for $\exp(-u/b)$ from (10.38)

$$\sum{}^* x_i \exp \frac{x_i}{b} \bigg/ \sum{}^* \exp \frac{x_i}{b} = b + \frac{1}{10} \sum x_i \tag{10.39}$$

From (10.39) we obtain by iteration $\hat{b} = 0.6705$ and, on substitution in (10.38), we find $\hat{u} = 0.8170$. [Note that usually these equations have to be solved simultaneously. In this particular case it so happened that we were able to derive one equation, namely, (10.39), involving only one unknown, namely, b, and another, namely, (10.38), expressing u in terms of b explicitly.]

Maximum likelihood estimators have certain desirable properties (e.g., consistency, minimum variance, etc.) These depend, however, on how realistically the model chosen represents the processes generating the data. Also, even when the model is valid, the desirable properties of the maximum likelihood estimators are assured only for sufficiently large sample sizes. In other words, the properties are asymptotic ones. Unfortunately, how large is "sufficiently large" is often unclear. Nevertheless, maximum likelihood estimators are good estimators and scholars tend to prefer them to other estimators in most practical situations.

Variances and Covariances of MLEs

To estimate the asymptotic variances and covariances of MLEs one first constructs the so-called *sample information matrix* (also known as the sample Hessian matrix) **I**, not to be confused with the notation for the identity matrix in matrix algebra. In the example given above, the sample information matrix is a 2×2 matrix

$$\begin{bmatrix} -\partial^2 L/\partial u^2 & -\partial^2 L/\partial b \partial u \\ -\partial^2 L/\partial u \partial b & -\partial^2 L/\partial^2 b \end{bmatrix}$$

where all the second-order partial derivatives are evaluated at the maximum likelihood estimates of u and b. From (10.36) and (10.37) it follows that

$$\frac{\partial^2 L}{\partial u^2} = -\left(\frac{1}{b}\right)^2 \sum{}^* \exp(W_i) \tag{10.40}$$

$$\frac{\partial^2 L}{\partial u \partial b} = \left(\frac{1}{b}\right)^2 \left[10 - \sum{}^* \exp(W_i) - \sum{}^* W_i \exp(W_i) \right] \tag{10.41}$$

and

$$\frac{\partial^2 L}{\partial b^2} = \left(\frac{1}{b}\right)^2 \left[10 + 2 \sum W_i - 2 \sum{}^* W_i \exp(W_i) - \sum{}^* W_i^2 \exp(W_i) \right] \quad (10.42)$$

so that the sample information matrix is

$$\begin{bmatrix} 26.45280 & 8.98139 \\ 8.98139 & 17.52283 \end{bmatrix} \quad (10.43a)$$

The inverse of this matrix, namely,

$$\begin{bmatrix} 0.0458 & -0.0235 \\ -0.0235 & 0.0691 \end{bmatrix} \quad (10.43b)$$

gives the variances and covariance of u and b. Thus the estimated standard error of u is $\sqrt{0.0458} = 0.2140$ and that of b is $\sqrt{0.0691} = 0.2629$.

Tests of Goodness of Fit

We mention here only the approximate likelihood ratio tests. Suppose the model has $(k + r)$ parameters. Let the parameters be $\alpha_1, \ldots, \alpha_k$ and $\beta_1, \ldots, \beta_r$. Let $L_u(\alpha, \beta)$ be the maximized likelihood when all the $(k + r)$ parameters are estimated. Suppose the null hypothesis specifies the values of the α parameters, and let $L_c(\alpha, \beta)$ be the maximized likelihood when only the β parameters are estimated, the α parameters being fixed at the hypothesized values. When the hypothesis is true, the statistic

$$X^2(k) = -2 \ln(L_c(\alpha, \beta)/L_u(\alpha, \beta)) \quad (10.44)$$

is approximately distributed as a chi square with k degrees of freedom. Large values of the statistic indicate poor fit between the hypothesis and the data.

4 Censoring

Survival analysis deals with the fact that for some individuals brought under observation the available information on failure time may be incomplete. In a life-testing experiment, for example, not all items may have failed by the close of the experiment. In a clinical trial, many patients survive to the end of the observation period. An individual under observation may migrate and hence or otherwise be lost to follow-up, before the observation period ends. (For an individual who is lost to follow-up after having been under observation, failure-free, for a length of time equal to t_c, say, we know only that his (her) failure time exceeds t_c). If for some case information on failure time is incomplete for such reasons as these, the case is said to provide *censored* information. Different censoring mechanisms are described in the literature. A few of these are mentioned below. For a thorough discussion of the topic, the reader is referred to Lawless (1982) and the literature cited therein.

Type I Censoring

If, in a life-testing experiment starting at time 0, n items are placed on test and the experiment is terminated at time t_c, then failure times will be known exactly only for those items that fail before t_c. In this setup all items have a constant potential censoring time, namely, t_c. Sometimes items differ in their potential censoring time. Thus suppose equipments installed at different dates are kept under observation until a fixed date t^*, say. In this setup, each equipment has its own potential censoring time (equal to the interval between installation date and the fixed date t^*.) When individuals (objects, items), $1, 2, \ldots, n$, are kept under observation for periods of lengths $c_1, c_2, \ldots, c_n$, respectively, so that the ith individual's failure time T_i is observed if and only if $T_i \leq c_i$, the resulting sample is said to be Type I censored.

The data from such a setup can be conveniently represented by n pairs of random variables (t_i, δ_i) where $t_i = \min(T_i, d_i)$ and

$$\delta_i = \begin{cases} 1, & \text{if } T_i \leq c_i, \quad \text{i.e., if the case is uncensored} \\ 0, & \text{if } T_i > c_i, \quad \text{i.e., if the case is censored.} \end{cases}$$

If, under this setup the T_i's are assumed to be independently and identically distributed random variables having continuous distributions with pdf $f(t)$ and survival function $S(t)$, then the likelihood on the data can be expressed as follows

$$\text{lik} = \prod_1^n [f(t_i)]^{\delta_i} [S(c_i)]^{1-\delta_i}$$

$$= \prod_u f(t_i) \prod_c S(c_i) \tag{10.45}$$

where the first product is over the uncensored cases and the second is over the censored cases.

Type II Censoring

If in a life-testing experiment, n items are placed on test and the experiment is terminated when a predetermined number of items have failed, the resulting sample is said to be Type II censored. A formal definition of Type II censoring can be given in terms of what are known as *order statistics*. Suppose n items, numbered $1, 2, \ldots, n$, are placed on test and their respective failure times are represented by $T_1, T_2, \ldots, T_n$. We use the notation $T_{(1)} \leq T_{(2)} \leq \cdots \leq T_{(n)}$ for the ordered failure times. Then omission of the $(n - r)$ greatest values $T_{(r+1)}, \ldots, T_{(n)}$ results in a Type II censored sample.

If $T_1, T_2, \ldots, T_n$ are independently and identically distributed random variables having continuous distribution with pdf $f(t)$ and survival function $S(t)$, then the likelihood of a sample resulting from censoring the $(n - r)$ greatest values is

$$f(t_{(1)}) \cdots f(t_{(r)})[S(t_{(r)})]^{n-r} \tag{10.46}$$

Random Censoring

In a clinical trial, patients usually enter the trial in a more or less random fashion, according to the date of diagnosis. If the observation period is terminated on a fixed date, then those patients who survive to the termination date are censored cases, the censoring time in a given case being the interval between the date of diagnosis and the date of termination. Survey data have similar characteristics. Consider for example data collected in an interview survey on the interval between marriage and first live birth. Individuals can be thought of as entering the study more or less randomly with respect to the date of marriage. Respondents with no live births as of the interview date are censored cases, the censoring time in any given case being the interval between marriage and the interview date. Ordinarily such data are treated as though they are Type I censored.

Other Forms of Censoring

Often the failure–censoring process is so complicated as to make precise modeling extremely difficult if not impossible. But experts seem to agree that the likelihood function given under Type I censoring can be used for purposes of drawing statistical inferences from the data in most practical situations (Lawless, 1982, p. 39).

Problems and Complements

1. If $t_1, t_2, \ldots, t_n$ are failure times presumed to be drawn from an exponential distribution with parameter α, show that the likelihood on the data is given by $\alpha^n \exp(-\alpha t_+)$, where $t_+ = t_1 + \cdots + t_n$. Show further that the maximum likelihood estimator of α is n/t_+.

2. Suppose in a Type II censored sample of size n, the failure times observed are $t_{(1)} < \cdots < t_{(r)}$, the value of r having been fixed before starting the data collection. From the general results on order statistics (see, e.g., Sarhan and Greenberg, 1962), show that under exponential sampling

the likelihood on the data is $\alpha^r \exp[-\alpha\{t_{(+)} + (n - r)t_{(r)}\}]$. The quantity within braces in the expression just given is sometimes referred to as the "total observed time." Show that the maximum likelihood estimate of α is r divided by total observed time. Does the maximum likelihood estimation procedure assume that all items fail sometime or other (e.g., marry, or have a child)? [ANSWER: No; all that is assumed is that the censored items survived until they were censored.]

3. In the setup of Exercise 2, suppose we write

$$w_1 = nt_{(1)};$$

and

$$w_i = (n - i + 1)[t_{(i)} - t_{(i-1)}], \qquad i = 2, \ldots, r$$

Verify that the total observed time is $w_1 + \cdots + w_r$, and that the Jacobian is

$$\frac{\partial(w_1, \ldots, w_r)}{\partial(t_{(1)}, \ldots, t_{(r)})} = \frac{n!}{(n - r)!}$$

Show that the joint pdf of $w_1, \ldots, w_r$ is $\alpha^r \exp(-\alpha w_+)$, where $w_+ = w_1 + \cdots + w_r$. What can we say about the distributions of the w's? Since the joint pdf of the w's is of the form $\Pi \; \alpha \exp(-\alpha w_i)$, where the product is over $i = 1, \ldots, n$, we infer that the w's are independently and identically distributed as exponential with parameter α.

4. In the setup of Exercise 3, show that the "total observed time"

$$V = t_{(1)} + \cdots + t_{(r)} + (n - r)t_{(r)}$$

is such that $2\alpha V$ is a chi square with $2r$ degrees of freedom. [HINT: The characteristic function of w_+ is $[1 - (i\theta)/\alpha]^{-r}$, which is the characteristic function of the gamma distribution with pdf

$$\frac{\alpha(\alpha t)^{r-1} \exp(-\alpha t)}{\Gamma(r)}$$

Therefore V is distributed as gamma with parameters α and r. A change of variable $2\alpha V = X$ shows that X has a chi-square distribution with $2r$ degrees of freedom.]

5. In the setup of Exercises 2 and 3, show that

$$t_{(r)} = \frac{w_1}{n} + \frac{w_2}{n - 1} + \cdots + \frac{w_r}{n - r + 1}$$

Show further that the expected value of $t_{(r)}$ is

$$\frac{1}{\alpha}\left[\frac{1}{n} + \frac{1}{n - 1} + \cdots + \frac{1}{n - r + 1}\right]$$

[HINT: The expected value of W with pdf $\alpha \exp(-\alpha w)$ is $1/\alpha$.]

6. If you were to plot the points $[t_{(i)}, m_i]$, where m_i is the sum of the reciprocals of $n, n - 1, \ldots,$ $n - i + 1$, and $t_{(1)} < \cdots < t_{(r)}$ are the r smallest observations from an exponential distribution, what pattern do you expect in the plot? [ANSWER: A linear pattern (see Exercise 5).]

7. Under the setup of Exercise 2, show that the product-limit estimate of the survival function [with $t_{(0)} = 0$ and $t_{(r+1)} = \infty$] for the interval $t_{(i)}$ to $t_{(i+1)}$ is

$$\prod_{j=1}^{i}\left[1 - \frac{1}{n - j + 1}\right]$$

the natural logarithm of which is approximately equal to the negative of the sum of the reciprocals of $n, n-1, \ldots, n-i+1$.

8. Based on your answers to Exercises 6 and 7, show that the plot of $t_{(i)}$ versus $\Sigma(n-j+1)^{-1}$, where the summation is from 1 to i, is a first-order approximation to the one based on $t_{(i)}$ versus the natural logarithms of the estimated survival function.

9. For the exponential with parameter α, the survival function being $S(t) = \exp(-\alpha t)$, what pattern do you expect for the plot of t versus $\ln(S(t))$?

10. Check whether the following failure times are from an exponential distribution:

$$1, \quad 1, \quad 2, \quad 2, \quad 3, \quad 4, \quad 4, \quad 5, \quad 6, \quad 8, \quad 8, \quad 9, \quad 10, \quad 10, \quad 12, \quad 14, \quad 16, \quad 20, \quad 24, \quad 34.$$

Recall that for the exponential distribution with parameter α, the cumulative distribution function is $F(t) = 1 - S(t) = 1 - \exp(-\alpha t)$. This implies that $t = -(1/\alpha)\ln[1 - F(t)]$. To check whether the given data are from an exponential, we may simply examine whether the regression of the observed t's on the empirically determined values of $\ln[1 - F(t)]$ is linear. It is common to estimate $F[t_{(i)}]$ as $(i - 0.5)/n$, where $t_{(i)}$'s are ordered observations in a random sample. Verify that the least-squares regression of $t_{(i)}$ on $\ln[21/(21 - i + 0.5)]$ in this case has an R^2 of 0.997, the intercept and slope of the regression line being 0.7606 and 8.8124, respectively. The parameter of the underlying exponential distribution can be estimated as the reciprocal of the 63rd percentile of the cumulative distribution, that is the reciprocal of $0.7606 + 8.8124 \ln(1/(1 - 0.632))$. These calculations can be done graphically using the exponential probability paper. We first calculate $F(t_{(i)}) = (i - 0.5)/n$ and plot the points whose x coordinates are $F(t_{(i)})$ and y coordinates are $t_{(i)}$ on the exponential probability paper, which has linear vertical scale and horizontal scale such that $F(t)$ is located at $\ln[1/1 - F(t)]$. After the points are plotted, one may fit a straight line by eye. [Show that for the exponential distribution with parameter α, the value of the cumulative distribution function $1 - \exp(-\alpha t)$ is 0.632 when $t = 1/\alpha$.]

11. The following are Type II censored data on failure times of certain airplane parts subjected to a life test (Mann and Fertig, 1973):

$$0.22, \quad 0.50, \quad 0.88, \quad 1.00, \quad 1.32, \quad 1.33, \quad 1.54, \quad 1.76, \quad 2.50, \quad 3.00$$

The sample size was 13. Check whether the data can be regarded as a Weibull sample. Recall that the Weibull distribution has the survival function $S(t) = \exp[-(\alpha t)^\beta]$. This implies that $\ln\{\ln[1/1 - F(t)]\} = \beta[\ln \alpha + \ln t]$. Fitting a straight line by ordinary least squares to the data points with $\ln\{\ln[1 - F(t))]\}$ and $\ln t$, respectively, as the x and y coordinates gives 0.8067 as the intercept and 0.7052 as the slope with an R^2 of 0.984. From the estimated slope and intercept, we obtain $\exp(-0.8067) = 0.446$ as our estimate of α and $1/0.7052 = 1.418$ as our estimate of β. The Weibull probability paper can be used for obtaining these results graphically. The probability paper converts $t_{(i)}$ to $\ln t_{(i)}$ and $F[t_{(i)}]$ to $\ln\{\ln[1/(1 - F(t_{(i)}))]\}$ automatically, the former on the horizontal axis and the latter on the vertical. When the data are Type II censored, only the points corresponding to the uncensored failure times are plotted. [The Weibull probability paper has an auxiliary scale that can be used in estimating the shape parameter β. A line is drawn parallel to the fitted line so as to pass through the origin of the auxiliary scale at the top of the paper, and then the point of intersection of that line with the vertical (auxiliary) scale on the left is determined. That point gives an estimate of β.]

12. Obtain a random sample from a Weibull distribution with 1.4 for the shape parameter and 0.5 for the location parameter. Use the relationship $\ln t = \ln(1/\alpha) + (1/\beta)\ln\{\ln[1/1 - F(t)]\}$, where β is the shape parameter and α the location parameter. Remember that $F(t)$ lies between 0 and 1. The following procedure thus suggests itself: Select a random number between 0 and 1.

Insert this number in place of $F(t)$ in the formula given above and compute $\ln t$. Exponentiation now gives a random failure time from the specified Weibull distribution.

13. If 3, 4, 5, 6, 8, 10 are ordered failure times from a Weibull distribution, estimate the parameters of the distribution using the method outlined in Exercise 11.

14. When the data are Type II censored, the observed failure times are the first k observations in a sample of size n, the number of censored cases being $(n - k)$. Let the ordered failure times observed be $t_{(1)} < \cdots < t_{(k)}$. The number of subjects at risk just prior to $t_{(j)}$ is $n_j = n - j + 1$. Show that in this case, an estimate of the value of the cumulative hazard function just after $t_{(j)}$ is the sum of the reciprocals of $n, n - 1, \ldots, (n - j + 1)$ [see Exercise 7].

15. Suppose the following are failure times (in seconds) of 20 insects in a study of effectiveness of an insecticide:

$$3, \quad 5, \quad 6, \quad 7, \quad 8, \quad 9, \quad 10, \quad 11, \quad 12, \quad 15, \quad 16, \quad 18, \quad 19, \quad 20, \quad 22, \quad 40, \quad 60$$

Check whether the data are from a log-normal distribution. Under the hypothesis that the given data are from a log-normal distribution $\ln t_{(i)}$ is linearly related to our estimate of the "plotting position" corresponding to $t_{(i)}$, i.e., the value of c such that the area to the left of c of a standard normal distribution is equal to $F(t_{(i)})$. Table 10.1 gives $t_{(i)}$, $(i - 0.5)/20$, and c_i such that the area to the left of c_i of a standard normal distribution is equal to $(i - 0.5)/20$. Treating the entries in the last column as predictor scores and $\ln t_i$ as response scores, we obtain by ordinary least squares a line with intercept equal to 2.65 and slope equal to 0.75, the corresponding R^2 being 0.994, indicating a good fit. Thus our estimates of the mean and the standard deviation of the log-normal distribution are 2.65 and 0.75, respectively.

Table 10.1

i	$t_{(i)}$	$(i - 0.5)/20$	c_i
1	3	0.025	-1.9600
2	5	0.075	-1.4395
3	6	0.125	-1.1503
4	7	0.175	-0.9346
5	8	0.225	-0.7554
6	9	0.275	-0.5978
7	10	0.325	-0.4538
8	11	0.375	-0.3186
9	12	0.425	-0.1891
10	15	0.475	-0.0627
11	16	0.525	0.0627
12	18	0.575	0.1891
13	19	0.625	0.3186
14	20	0.675	0.4538
15	22	0.725	0.5978
16	25	0.775	0.7554
17	28	0.825	0.9346
18	30	0.875	1.1503
19	40	0.925	1.4395
20	60	0.975	1.9600

16. The following data show the number of million revolutions before failure for each of 23 ball bearings in a life test (Lieblein and Zelen, 1956):

17.88, 28.92, 33.00, 41.52, 42.12, 45.60, 48.40, 51.84, 51.96, 54.12, 55.56

67.80, 68.64, 68.64, 68.88, 84.12, 93.12, 98.64, 105.12, 105.84, 127.92, 128.04, 173.40

Check whether the data can be regarded as a log-normal sample.

17. Suppose $t_{(1)} < \cdots < t_{(n)}$ are failure times. Regard the data as an uncensored gamma sample, and outline the procedure for estimating the parameters of the underlying gamma distribution. See Lawless (1982). [The pdf of gamma distribution is $[1/\Gamma(\beta)]\alpha(\alpha t)^{\beta-1} \exp(-\alpha t)$.]

18. Apply the procedure you have outlined in Exercise 17 to the following hypothetical data: 5, 8, 10, 12, 16, 22, 23, 30, 42. Let R be the ratio of the geometric mean of the observations to their arithmetic mean. Wilk *et al.* (1962) have prepared a table that gives estimate of β in terms of $1/(1 - R)$. An estimate of α can then be obtained as the product of the estimate of β obtained from the table and $n/\Sigma\, t_{(i)}$. For the data in question we thus get 3.024 as our estimate of β and 0.165 as our estimate of α. [Interpolation is necessary to estimate β.] These estimates are known to be biased. Lee (1980) refers to the following correction for bias: Divide the figure obtained from the Wilk *et al.* (1962) table by $[(3/n) + 1]$ to obtain a corrected estimate of β. Multiply this corrected estimate of β by the product

$$[n/\sum t_{(i)}][1 - 1/(n \times \text{corrected estimate of } \beta)]$$

to obtain a corrected estimate of α. For the data in question, the corrected estimates of β and α thus obtained are 2.326 and 0.118, respectively.

19. When censoring is present, more complicated calculations become necessary (see Wilk *et al.* 1962).

20. A regression method for fitting hazard functions has been suggested by Gehan and Siddiqui (1973) for samples from exponential, Weibull, Gompertz, and linear exponential distributions.

Bibliographic Notes

Elandt–Johnson and Johnson (1980) give an excellent modern account of survival measurements and concepts. For details regarding individual distributions refer to Lawless (1982), Gross and Clark (1975), and the four-volume encyclopedic work on distributions by Johnson and Kotz (1969–1972).

Chapter 11 | Exponential, Piecewise Exponential, and General Linear Models

It is assumed that the reader is familiar with the use of regression models in the analysis of experimental and nonexperimental data. In the statistical theory of linear models, it may be recalled, the response variable in the experimental setting is assumed to be measured when all the factor variables are controlled. In the nonexperimental setting, where controlling the factors is impossible, the factors are assumed to be "fixed" with respect to the response variable. The model, e.g., $y = \beta_0 x_0 + \beta_1 x_1 + \cdots + \beta_k x_k + \epsilon$, is assumed to be correct. The residuals (ϵ's) are assumed to have expectation zero, and are uncorrelated across observations and have common variance. For testing, it is further assumed that the ϵ's are normally distributed. Under the assumptions, it is known that the least-squares technique provides the best linear unbiased estimates of the β's. If the residuals are normally distributed, it is also known that under certain hypotheses, each of the following holds: (1) Various sums of squares are distributed proportional to chi square, (2) Ratios of estimates to their respective standard errors are distributed as student's t, (3) Appropriate ratios of sums of squares are distributed as F. The question arises whether survival data, with covariates, can be analyzed using the usual regression approach just referred to. Two peculiarities of survival data demand attention in this connection: (1) nonnormality and (2) censoring. In this chapter, we shall discuss how these features are usually handled when fitting regression models to survival data.

1 Sampling from the Exponential Distribution

The Single-Sample Problem

To prepare a background, we start with a discussion of the single-sample problem. The focus is on sampling from the exponential. Suppose the following sample of failure times is assumed to have been drawn from an exponential distribution with pdf $f(t; \alpha) = \alpha \exp(-\alpha t)$:

$$32, \quad 61, \quad 149, \quad 190, \quad 307, \quad 472, \quad 500, \quad 681$$

Suppose we wish to test whether the mean of the exponential distribution from which the sample was drawn is 250. We shall use the likelihood ratio test. The likelihood on the data for α is

$$\text{lik} = f(32; \alpha) f(61; \alpha) \cdots f(681; \alpha)$$

$$= \alpha^8 \exp[-(32 + \cdots + 681)\alpha]$$

$$= \alpha^8 \exp(-2392\alpha) \tag{11.1}$$

The log likelihood is

$$L(\alpha) = 8 \ln \alpha - 2392\alpha \tag{11.2}$$

and

$$dL(\alpha)/d\alpha = (8/\alpha) - 2392 \tag{11.3}$$

Therefore the maximum likelihood estimate of α, obtained by solving for α from $dL(\alpha)/d\alpha = 0$, is $\hat{\alpha} = \frac{8}{2392} = \frac{1}{299}$. The corresponding estimate of the mean of the distribution is 299. To test whether the mean is 250, we test whether $\alpha = \frac{1}{250}$. The maximized log likelihood is

$$L_{\text{max}} = 8 \ln\frac{1}{299} - 2392 \times \frac{1}{299} = -53.60355$$

The value of the log likelihood when α takes the hypothesized value $\frac{1}{250}$ is

$$L_{\text{H}} = 8 \ln\frac{1}{250} - 2392 \times \frac{1}{250} = -53.73969$$

The likelihood ratio statistic, for testing the hypothesis H: $\alpha = \frac{1}{250}$,

$$X^2(1) = -2[L_{\text{H}} - L_{\text{max}}] = 0.273$$

turns out to be too small (compared to 3.84, the 5 percent value of the chi-square distribution with 1 degree of freedom) to reject the hypothesis. In general, if the sample $\{t_1, t_2, \ldots, t_n\}$ is assumed to have come from an exponential distribution with pdf $f(t; \alpha) = \alpha \exp(-\alpha t)$, the likelihood on the data for α is

$$\text{lik} = \prod_1^n f(t_i; \alpha) = \alpha^n \exp\left(-\sum t_i \alpha\right) \tag{11.4}$$

The log likelihood is

$$L(\alpha) = n \ln \alpha - \alpha\left(\sum t_i\right)$$

so that the maximum likelihood estimate of α is given by the solution to the equation

$$dL(\alpha)/d\alpha = n/\alpha - \sum t_i = 0 \tag{11.5}$$

The solution is easily seen to be

$$\hat{\alpha} = n \Big/ \sum t_i \tag{11.6}$$

from which it is clear that the sample mean is an estimate of $1/\alpha$. To test the null hypothesis $H_0: \alpha = \alpha_0$ we use the likelihood ratio statistic

$$X^2(1) = -2[(\text{log likelihood when } \alpha = \alpha_0) - (\text{log likelihood when } \alpha = \hat{\alpha})]$$

which, when the hypothesis H_0 is true, can be treated as a chi square with 1 degree of freedom.

An Alternative Way of Testing $H_0: \alpha = \alpha_0$

From the expression for the log likelihood $L(\alpha)$, we have already seen that $dL(\alpha)/d\alpha = n/\alpha - \sum t_i$. Differentiation gives $d^2L(\alpha)/d\alpha^2 = -n/\alpha^2$, so that α^2/n, i.e., the negative of the reciprocal of $d^2L(\alpha)/d\alpha^2$, evaluated at $\alpha = \hat{\alpha}$, where $\hat{\alpha}$ is the maximum likelihood estimate of α, is an estimate of the variance of α. In large samples, when the hypothesis H_0 is true,

$$(\hat{\alpha} - \alpha) \Big/ \sqrt{\sum \frac{\alpha^2}{n}} \tag{11.7}$$

is distributed as the standard normal. In the numerical example given above, $\hat{\alpha} = \frac{1}{299}$, and $\alpha = \frac{1}{250}$. Therefore, the corresponding values of the statistic (11.7) is -0.385, which is too small to reject H_0.

Single sample with censoring. Suppose starting at time 0 three items were placed on test and that two of them failed at times t_1 and t_2, respectively, while the remaining one was censored at time t_c. Then if we assume that the sample has been drawn from an exponential distribution with parameter α, the likelihood on the data for α is

$$f(t_1; \alpha) f(t_2; \alpha) S(t_c; \alpha)$$

or

$$[\alpha \exp(-\alpha t_1)][\alpha \exp(-\alpha t_2)][\exp(-\alpha t_c)]$$

which simplifies to

$$\alpha^2 \exp(-\alpha t_+)$$

where $t_+ = t_1 + t_2 + t_c$, the total observed exposure time.

In general, if $t_{f1}, t_{f2}, \ldots, t_{fm}$ are m failure times and $t_{c1}, \ldots, t_{cn}$ are n censoring times, all assumed to be drawn from an exponential distribution with parameter α, then the likelihood on the data for α is

$$\text{lik} = \alpha^m \exp(-\alpha t_+)$$

where $t_+ = (t_{f1} + \cdots + t_{fm}) + (t_{c1} + \cdots + t_{cn})$ and the log likelihood is

$$L(\alpha) = m \ln \alpha - \alpha t_+ \qquad (11.8)$$

from which we obtain

$$dL(\alpha)/d\alpha = (m/\alpha) - t_+ \qquad (11.9)$$

yielding

$$\hat{\alpha} = m/t_+ \qquad (11.10)$$

as the maximum likelihood estimate of α.

Example. Bartholomew (1957) reports the following data on failure and censoring times observed in a life-testing experiment involving an industrial equipment:

$$2, \quad 4, \quad 14, \quad 21^*, \quad 24, \quad 27, \quad 33, \quad 51, \quad 60^*, \quad 71^*$$

where * indicates censoring. The total observed exposure time is $2 + 4 + \cdots + 71 = 307$, and the number of failures is 7. If we assume that the data were drawn independently from an exponential with parameter α, the maximum likelihood estimate of α is $\frac{7}{307} = 0.022801$.

Testing Hypotheses

Several methods are available to test hypotheses concerning the parameter α in the context described above.

1. One method is to use the asymptotic normality of the maximum likelihood estimators. In the present context $\hat{\alpha}$ is asymptotically normally distributed with mean α and estimated variance given by the negative of the reciprocal of $d^2L(\alpha)/d\alpha^2$, evaluated at $\alpha = \hat{\alpha}$.
2. A second method is to use the result demonstrated by Epstein and Sobel (1957) that $\ln \hat{\alpha}$ is normally distributed with mean $\ln \alpha$ and variance $1/m$, where m is the number of observed failures. [Notice that the variance of $\ln \hat{\alpha}$ is independent of α. Because of this property, the distribution of $\ln \hat{\alpha}$ approaches the normal faster than does that of $\hat{\alpha}$. The implication is that the Epstein–Sobel method may be used for smaller samples.]

3. Yet another method is that of Sprott (1973), according to which one may use in small samples the statistic $\phi = \hat{\alpha}^{1/3}$, which is approximately normal with mean $\alpha^{1/3}$ and estimated variance $\hat{\alpha}^{2/3}/9m$.

4. One may also use the result that the likelihood ratio statistic, $-2(L_H - L_{max})$, where L_{max} is the maximum value of the log likelihood, and L_H is the value of the log likelihood evaluated at the hypothesized value of α, is distributed in large samples as a chi square with 1 degree of freedom. For Bartholomew's (1957) data we obtained the maximum likelihood estimate $\hat{\alpha} = 0.022801$. Using the data, let us compute 95 percent confidence intervals for α by methods (2), (3), and (4) mentioned above.

The Epstein–Sobel method gives the following 95 percent confidence interval for $\ln \alpha$:

$$\ln \hat{\alpha} - 1.96/\sqrt{m} \leq \ln \alpha \leq \ln \hat{\alpha} + 1.96/\sqrt{m} \tag{11.11}$$

from which we get the following 95 percent confidence interval for α, remembering that $m = 7$ and $\hat{\alpha} = 0.022801$:

$$0.01087 \leq \alpha \leq 0.047828$$

or

$$20.91 \leq 1/\alpha \leq 92.00$$

for $1/\alpha$ the mean of the distribution.

The method of Sprott (1973) gives the following 95 percent confidence interval for $\alpha^{1/3}$:

$$\hat{\alpha}^{1/3} - 1.96\hat{\alpha}^{1/3}/(3\sqrt{m}) \leq \alpha^{1/3} \leq \hat{\alpha}^{1/3} + 1.96\hat{\alpha}^{1/3}/(3\sqrt{m}) \tag{11.12}$$

from which we get the following 95 percent confidence interval for α:

$$0.0097376 \leq \alpha \leq 0.044206$$

or

$$22.02 \leq (1/\alpha) \leq 102.70$$

for the mean of the distribution.

To apply the likelihood ratio statistic method, we notice that

$$L_{max} = m \ln \hat{\alpha} - \hat{\alpha}t_+$$

and that

$$L = m \ln \alpha - \alpha t_+$$

Hence the likelihood ratio statistic has the form

$$-2[m \ln \alpha - \alpha t_+ - m \ln \hat{\alpha} + \hat{\alpha}t_+]$$

which is equivalent to the following:

$$-2m[\ln(\alpha/\hat{\alpha}) - (\alpha/\hat{\alpha}) + 1]$$

in obtaining which we use the estimating equation (11.10) for α obtained earlier, namely, $\hat{\alpha}t_+ = m$, to replace αt_+ by $(\alpha/\hat{\alpha})m$ and $\hat{\alpha}t_+$ by m. Now remembering that the likelihood ratio statistic in the present case is asymptotically distributed as a chi square, we solve the following equation for $(\alpha/\hat{\alpha})$:

$$3.84 = -2m[\ln(\alpha/\hat{\alpha}) - \alpha/\hat{\alpha}) + 1] \qquad (11.13)$$

where 3.84 is the 5 percent point of the chi square with 1 degree of freedom. This last equation can be written as

$$y - \ln y = 1.274286 \qquad (11.14)$$

where $y = \alpha/\hat{\alpha}$. We should expect two values of y to satisfy the equation given above, since $y - \ln y$ monotonically decreases to 1 as y moves to 1 from the left and then monotonically increases as y moves to the right from 1. By trial we find that 0.4298 and 1.9337 are the solutions for y we are looking for. These in turn lead to the following 95 percent confidence interval for α:

$$0.0097999 \leq \alpha \leq 0.04409$$

or

$$22.68 \leq 1/\alpha \leq 102.04$$

for $1/\alpha$.

Notice that the three methods yield more or less the same confidence intervals for $1/\alpha$. There seems to be consensus that these methods can be used for small samples (Lawless, 1982).

2 Piecewise Exponential

So far we have assumed that a single exponential distribution applies to the entire time (age) range. We now entertain the possibility that this assumption may not be valid. Thus, the hazard rate characterizing mortality risks due to cancer is known to vary with duration since diagnosis. Let us first consider the simple case in which the hazard rate remains constant at one level (say, α_1) until time (age) y_1, and at another level (say, α_2) from time (age) y_2 onward. That is,

$$h(t) = \begin{cases} \alpha_1 & \text{for} \quad 0 < t < y_1 \\ \alpha_2 & \text{for} \quad y_1 \leq t < \infty \end{cases}$$

Then clearly

$$f(t) = \begin{cases} \alpha_1 \exp(-\alpha_1 t) & \text{for} \quad 0 \le t \le y_1 \\ \alpha_2 \exp(-\alpha_1 y_1) \exp[-\alpha_2(t - y_1)] & \text{for} \quad y_1 \le t < \infty \end{cases}$$

and

$$S(t) = \begin{cases} \exp(-\alpha_1 t) & \text{for} \quad 0 \le t < y_1 \\ \exp(-\alpha_1 y_1) \exp[-\alpha_2(t - y_1)] & \text{for} \quad y_1 \le t < \infty \end{cases}$$

Now let A, B, C, and D be four individuals such that

$$\begin{array}{lll} \text{A} & \text{failed at} & t_A < y_1 \\ \text{B} & \text{was censored at} & t_B < y_1 \\ \text{C} & \text{failed at} & t_C > y_1 \end{array}$$

and

$$\begin{array}{lll} \text{D} & \text{failed at} & t_D > y_1 \end{array}$$

Then it is easy to see that the contributions of these individuals to the likelihood are

$$\alpha_1 \exp(-\alpha_1 t_A) \quad \text{from A}$$

$$\exp(-\alpha_1 t_B) \quad \text{from B}$$

$$\alpha_2 \exp(-\alpha_1 y_1) \exp[-\alpha_2(t_C - y_1)] \quad \text{from C}$$

and

$$\exp(\alpha_1 y_1) \exp[-\alpha_2(t_D - y_1)] \quad \text{from D}$$

The combined contribution of these four individuals to the likelihood is the product of their individual contributions just mentioned. Their combined contribution to the log likelihood is then given by

$$L = \ln \alpha_1 + \ln \alpha_2 - \alpha_1(t_A + t_B + 2y_1) - \alpha_2[(t_C - y_1) + (t_D - y_1)]$$

$$= \ln \alpha_1 + \ln \alpha_2 - \alpha_1 t_{1+} - \alpha_2 t_{2+} \tag{11.15}$$

where $t_{1+} = t_A + t_B + y_1 + y_1 = $ the total observed exposure time in the first segment of the time axis, i.e., in $[0, y_1)$, and $t_{2+} = (t_C - y_1) + (t_D - y_1) = $ the total observed exposure in the second segment of the time axis, i.e., in $[y_1, \infty)$. Generalizing from this example, if n_1 failures are observed during the first segment and n_2 during the second segment, and if the total observed exposure times in these segments are t_{1+} and t_{2+}, respectively, then the log likelihood on the data is

$$L = n_1 \ln \alpha_1 + n_2 \ln \alpha_2 - \alpha_1 t_{1+} - \alpha_2 t_{2+}$$

$$= \sum_{i=1}^{2} (n_i \ln \alpha_i - \alpha_i t_{i+})$$

In general, if the time axis is segmented into I subintervals, with the ith subinterval characterized by a constant hazard rate equal to α_i, say, and if the data show n_i failures and t_{i+} to be the total observed exposure time in the ith subinterval, then the log likelihood on the data for $\alpha_1, \alpha_2, \ldots, \alpha_I$ is

$$L = \sum_{i=1}^{I} (n_i \ln \alpha_i - \alpha_i t_{i+}) \tag{11.16}$$

Clearly, from (11.16), for $i = 1, \ldots, I$,

$$\partial L(\alpha_1, \ldots, \alpha_I)/\partial \alpha_i = (n_i/\alpha_i) - t_{i+} \tag{11.17}$$

The maximum likelihood estimate of α_i is therefore given by

$$\hat{\alpha}_i = n_i/t_{i+}, \qquad i = 1, \ldots, I \tag{11.18}$$

We may call this estimate descriptively the "observed risk in the ith time interval." Notice that the log likelihood (11.16) is, up to an additive constant, the same as

$$\sum [n_i \ln(\alpha_i t_{i+}) - \alpha_i(t_{i+})] \tag{11.19}$$

provided that we treat t_{i+} as fixed for all i.

Now let us consider each n_i as Poisson, conditional on t_{i+}, with

$$E(n_i \mid t_{i+}) = \alpha_i t_{i+}, \qquad i = 1, 2, \ldots, I \tag{11.20}$$

[The random variable X is Poisson with parameter α if

$$\Pr(X = x) = (\alpha^x/x!) \exp(-\alpha)$$

for $x = 0, 1, 2, \ldots$. It is easy to see that $E(X) = \alpha$.] Under this setup, the log likelihood is, to within an additive constant

$$\sum [n_i \ln(\alpha_i t_{i+}) - \alpha_i t_{i+}] \tag{11.21}$$

which is the same as (11.16). Since the two sampling schemes give log likelihoods that differ utmost by an additive constant, we can consider the two log likelihoods as interchangeable for purposes of drawing inferences based on maximum likelihood estimators and/or likelihood ratio statistics. At this point it is useful to note that the Poisson with parameter α can be thought of as the limit of a sequence of binomials, as the number of independent trials increases to infinity and the probability of a particular outcome of interest, e.g., failure, decreases to zero, while the product of the two remains constant at α (Feller, 1968). This implies that for sufficiently large t_{i+} and sufficiently small α, we may replace the I independent Poissons n_i, conditional on t_{i+}, having conditional expectations, given t_{i+}, equal to $\alpha_i t_{i+}$, with I independent binomials n_i, conditional on t_{i+}, having conditional expectations, given t_{i+}, equal to $\alpha_i t_{i+}$. The data viewed as I independent binomials may be displayed

Table 11.1

Piecewise Exponential Samples Viewed as a Collection of Independent
Binomial Samples

Observed number of failures	Observed number of nonfailures	Total	Probability of failure
$x_{11} = n_1$	x_{12}	$x_{1+} = t_{1+}$	α_1
$\vdots$	$\vdots$	$\vdots$	$\vdots$
$x_{I1} = n_I$	x_{I2}	$x_{I+} = t_{I+}$	α_I

in the form of a two-way table as shown in Table 11.1. For the purpose of computing log likelihood, we may equivalently regard the x_{ij}'s displayed in Table 11.1 as a multinomial having $2I$ categories (I time intervals, two outcomes) subject to the constraints that each row total be treated as fixed (Bishop *et al.*, 1975, chapter 13). Put differently, for purposes of deriving log likelihoods, we may regard the x_{ij}'s in Table 11.1 as a two-way classification with fixed row marginals. Notice that this way of looking at the data is merely a device to derive log likelihoods; there is no need to assume that the data were in fact generated in the fashion implied. The advantage of viewing the data as a multinomial with a set of marginals fixed is that we can, under the setup, use for the analysis of the data the theory, methods, and computer programs developed for the analysis of contingency tables. The two-sample case discussed below illustrates the point.

Two Sample Problems

Suppose we have data from two samples, as displayed in Table 11.2. We assume that both samples are drawn from piecewise exponentials: Sample 1 from exponentials with parameters $\alpha_{11}, \ldots, \alpha_{I1}$, the number of failures and total exposure time observed under the exponential with parameter α_{i1} being n_{i1} and t_{i1+}, respectively, $i = 1, \ldots, I$; and Sample 2 from exponentials with

Table 11.2

Data Structure of a Two Sample Problem

Time interval	Sample 1		Sample 2	
	Outcome 1 (failure)	Outcome 2 (nonfailure)	Outcome 1 (failure)	Outcome 2 (nonfailure)
1	$x_{111} = n_{11}$	$x_{112} = t_{11+} - n_{11}$	$x_{121} = n_{12}$	$x_{122} = t_{12+} - n_{12}$
$\vdots$	$\vdots$	$\vdots$	$\vdots$	$\vdots$
I	$x_{I11} = n_{I1}$	$x_{I12} = t_{I1+} - n_{I1}$	$x_{I21} = n_{I2}$	$x_{I22} = t_{I2+} - n_{I2}$

parameters $\alpha_{12}, \ldots, \alpha_{I2}$, the number of failures and the total exposure time observed under the one with parameter α_{i2} being n_{i2} and t_{i2+}, respectively.

Suppose we wish to test the hypothesis that the two samples have proportional hazards, that is, that α_{i1}/α_{i2} is invariant over i.

To test this hypothesis, we regard the data in Table 11.2 as a multinomial sample with $I \times 2 \times 2$ categories (I time intervals, two samples, and two outcomes) with $I \times 2$ constraints: within each time interval of each sample, the sum of the expected values of the number of failures and the number of "nonfailures" equals the observed total exposure time, i.e., in the notation of Table 11.2

$$E(x_{ij1} + x_{ij2}) = x_{ij+}, \qquad i = 1, \ldots, I; \qquad j = 1, 2 \qquad (11.22)$$

Let us denote the expected value of x_{ijk} by m_{ijk} and let

$$m_{ijk} = \theta \theta_i^A \theta_j^S \theta_k^O \theta_{ij}^{AS} \theta_{ik}^{AO} \theta_{jk}^{SO} \theta_{ijk}^{ASO}$$

where A stands for age (time), S for sample, and O for outcome (failure, nonfailure.) [Some readers may find the notation used here confusing. The notation involving superscripts and subscripts has become more or less standard in the literature on categorical data analysis. One way to become acquainted with the notation is to think in terms of cells in one-way tables, two-way tables, and so on. The superscripts indicate the particular variables involved in the tables, and the subscripts indicate the typical cells. Thus when AS appears in the superscript, think of the two-way table formed by crossing A with S; when AS appears as the superscript and ij as the subscript, think of the (i, j) cell of the A by S table; and so on.]

In order that there be no more parameters than there are "observations" (cells in the table), we introduce the following constraints on the θ parameters:

$$\prod_i \theta_i^A = \prod_j \theta_j^S = \prod_k \theta_k^O = 1$$

$$\prod_i \theta_{ij}^{AS} = \prod_j \theta_{ij}^{AS} = 1$$

$$\prod_i \theta_{ik}^{AO} = \prod_k \theta_{ik}^{AO} = 1$$

$$\prod_j \theta_{jk}^{SO} = \prod_k \theta_{jk}^{SO} = 1$$

$$\prod_i \theta_{ijk}^{ASO} = \prod_j \theta_{ijk}^{ASO} = \prod_k \theta_{ijk}^{ASO} = 1$$

Notice that with these constraints, the number of free θ's is equal to $I \times 2 \times 2$:

Superscript of θ	None	A	S	O	AS	AO	SO	ASO
Number of such θ's free	1	$I - 1$	1	1	$I - 1$	$I - 1$	1	$I - 1$

The hypothesis of proportional hazards we wish to test needs to be specified in terms of the θ parameters. First, notice that the hypothesis is the same as the following:

$$\frac{m_{i11}/m_{i1+}}{m_{i21}/m_{i2+}} \quad \text{is invariant over} \quad i \tag{11.23}$$

When m_{ijk}/m_{ij+} is very small, (11.23) is approximately the same as saying that

$$\frac{m_{i11}/m_{i12}}{m_{i21}/m_{i22}} \quad \text{is invariant over} \quad i \tag{11.24}$$

But, as can be easily verified,

$$\frac{m_{i11}m_{i22}}{m_{i12}m_{i21}} = \frac{\theta_{11}^{SO}\theta_{i11}^{ASO}\theta_{22}^{SO}\theta_{i22}^{ASO}}{\theta_{12}^{SO}\theta_{i12}^{ASO}\theta_{21}^{SO}\theta_{i21}^{ASO}} \tag{11.25}$$

the right-hand side of which is invariant over i if all three-subscripted parameters are unity. To fit the model with all parameters with three subscripts set equal to unity, we fit the three two-way marginals; that is, we compute cell entries that reproduce the observed two-way marginals. The cell entries thus computed are the estimated expected cell entries; i.e., estimates of m_{ijk}'s. They satisfy the following equations (using the notation in Table 2):

$$m_{ij+} = x_{ij+} \tag{11.26}$$

$$m_{i+k} = x_{i+k} \tag{11.27}$$

$$m_{+jk} = x_{+jk} \tag{11.28}$$

which incidentally include the constraints introduced earlier, namely, that the expected values of x_{ij1} and x_{ij2} must sum to x_{ij+}. One uses iterative proportional fitting algorithms to obtain cell entries that satisfy equations (11.26), (11.27), and (11.28). The estimated cell entries are then compared to the observed entries to determine goodness of fit. Existing computer programs such as those in the "BMPD Statistical Software, 1981" (Dixon et al., 1981) can be readily used for the purpose (see the following example). It is easily seen that there is no new principle involved if (see Table 11.2) there are three or more samples to be compared. If the samples themselves are defined in terms of combinations of categories of covariates, then the data on hand can be treated as a multiway contingency table for purposes of analysis. The following example makes this clear.

Example. Laird and Olivier (1981) analyzed the data shown in Table 11.3 using the method outlined above. They considered two covariates: donor type, a dichotomy (cadaveric, nonrelated; living, related); and match grade, a five-point ordinal variable (0, 1, 2, 3, 4, 5) based on the number of matched antigens.

Table 11.3

Data on Graft-Survival Following Kidney Transplant Operations [M = Match Grade; R = Donor Relationship][a,b]

Survival time d (days) m (months) y (years)	M = 0			M = 1			M = 2			M = 3			M = 4		
	E	W	D	E	W	D	E	W	D	E	W	D	E	W	D
R = 1 (cadaveric, nonrelated)															
0– 7 d	242	0	13	386	0	16	393	0	20	130	0	8	18	0	0
7– 15 d	229	0	18	370	0	25	373	0	20	122	0	4	18	0	1
15– 21 d	211	0	14	345	0	12	353	0	18	118	0	8	17	0	0
21– 30 d	197	0	13	333	0	17	335	0	18	110	0	10	17	0	0
1– 2 m	184	0	21	316	0	43	317	0	36	100	0	14	17	0	1
2– 3 m	163	0	14	273	0	30	281	0	26	86	0	2	16	0	3
3– 6 m	149	0	23	243	0	35	255	0	36	84	0	10	13	0	2
6– 9 m	126	0	8	208	0	11	219	0	6	74	0	5	11	0	0
9– 12 m	118	0	3	197	0	16	213	0	10	69	0	1	11	0	0
1–1.5 y	115	0	4	181	0	9	203	0	11	68	0	3	11	0	0
1.5– 2 y	111	2	9	172	6	6	192	8	7	65	2	4	11	0	1
2–2.5 y	100	17	2	160	34	8	177	43	8	59	14	0	10	5	1
2.5– 3 y	81	21	3	118	26	4	126	31	6	45	11	0	4	1	1
3–3.5 y	57	19	2	88	27	3	89	22	5	34	11	1	2	0	0
3.5– 4 y	36	19	1	58	33	1	62	27	1	22	11	0	2	0	0
>4 y	16	16	0	24	23	1	34	34	0	11	9	2	2	2	0
R = 2 (living, related)															
0– 7 d	20	0	0	48	0	2	410	0	14	164	0	4	164	0	1
7– 15 d	20	0	1	46	0	0	396	0	7	160	0	3	163	0	2
15– 21 d	19	0	1	46	0	1	389	0	3	157	0	5	161	0	1
21– 30 d	18	0	0	45	0	2	386	0	4	152	0	7	160	0	1
1– 2 m	18	0	3	43	0	5	382	0	31	145	0	8	159	0	5
2– 3 m	15	0	1	38	0	3	351	0	21	137	0	1	154	0	1
3– 6 m	14	0	2	35	0	2	330	0	27	136	0	8	153	0	4
6– 9 m	12	0	0	33	0	0	303	0	14	128	0	3	149	0	1
9– 12 m	12	0	0	33	0	1	289	0	11	125	0	1	148	0	2
1–1.5 y	12	0	0	32	0	2	279	0	9	124	0	0	146	0	0
1.5– 2 y	12	0	0	30	0	0	269	8	10	124	2	1	146	4	2
2–2.5 y	12	2	0	30	4	0	251	53	7	121	17	2	140	23	1
2.5– 3 y	10	0	0	26	8	1	191	45	3	102	17	1	116	33	1
3–3.5 y	10	5	0	17	3	0	143	45	2	84	26	1	82	27	1
3.5– 4 y	5	1	0	14	7	0	96	39	0	57	24	0	54	23	1
>4 y	4	4	0	7	7	0	57	57	0	33	33	0	30	30	0

[a] Source: Laird and Olivier (1981).
[b] Under each match grade (M), the first column (E) gives the number entered, the next (W), the number withdrawn, and the last (D), the number of deaths.

Survival time was grouped into 16 categories of varying lengths. As can be seen from Table 11.3, the preliminary data set was in the form of 10 life tables, showing for each time interval, under each combination of match grade and donor type, the number entered E, the number died D, and the number withdrawn W. For each time interval, the total observed exposure time was calculated by multiplying $[E - 0.5(W + D)]$ by the width of the interval. Thus for the first time interval for those with donor type 1 and match grade 0, the total observed exposure time was estimated as $7 \times [242 - 0.5(0 + 13)] =$ 1648.5. For these calculations, the last open interval was taken as having a width of 180 days. The total observed exposure times thus calculated are shown in Table 11.4, along with the number of deaths in each interval (reproduced from Table 11.3). The following hierarchical model was fitted to the data thus created: $[RSM, RO, MO, SO]$, where R stands for donor type, taking values 1 if cadaveric, nonrelated, and 2 if living, related; M for match grade, taking values 0, 1, 2, 3, 4, or 5; S for survival time, with 16 categories, and O for outcome (died, not died, the "count" under the latter being equal to the total observed exposure time minus the number died). Note that the model is specified in terms of the marginals fitted. The variable O is viewed as the response variable, and R, M, and S as factors. The inclusion of RMS among the marginals fitted amounts to "fixing" the observed total exposure time in each factor combination (donor type by match grade by time interval.) For model fitting, Laird and Olivier (1981) used a program called LOGLIN, produced by the Health Sciences Computing Facility at the Harvard School of Public Health. We used the program for multiway contingency table analysis in $BMDP$ (Dixon et al., 1981), and obtained a likelihood ratio statistic of 152.2 with 139 degrees of freedom, indicating a reasonably good fit. (Laird and Olivier obtained a likelihood ratio statistic of 154.8 with 139 degrees of freedom.)

The $BMDP$ computer output permits estimating the relative odds on failure. For match grade, the relative odds are, treating grade 0 as the reference category: 1.00, 0.97, 0.76, and 0.33 for grades 1, 2, 3, and 4, respectively. The odds on failure decrease appreciably only after the match grade improves beyond 2. The relative odds on failure, compared to those of patients with match grade 0, are only three-fourths and one-third, respectively, for patients with match grades 3 and 4. As for the impact of donor type, patients receiving the organ (kidney) from living relatives have only one-half as much odds on failure as have those receiving the organ from nonrelated cadavers. Finally, the odds on failure sharply diminish with duration since transplant, as the following pattern of relative odds, using the first time interval, 0 to 7 days, as the reference category shows:

0.95,   1.06,   0.84,   0.60,   0.41,   0.22,   0.08,   0.08

0.01,   0.01,   0.02,   0.03,   0.06,   0.14,   0.22

Table 11.4

Number of Deaths and Exposure Time Computed from the Data in Table 11.3[a]

Survival time (S)	M = 0		M = 1		M = 2		M = 3		M = 4	
	D	T	D	T	D	T	D	T	D	T
R = 1 (cadaveric, unrelated)										
1	13	1648.5	16	2646.0	20	2681.0	8	882.0	0	126.0
2	18	1760.0	25	2860.0	20	2940.0	4	960.0	1	140.0
3	14	1224.0	12	2034.0	18	2064.0	9	684.0	0	102.0
4	13	1714.5	17	2920.5	18	2934.0	10	945.0	0	153.0
5	21	5205.0	43	8835.0	36	8970.0	14	2790.0	1	495.0
6	14	4680.0	30	7740.0	26	8040.0	2	2550.0	3	435.0
7	23	12375.0	35	20295.0	36	21330.0	10	7110.0	2	1080.0
8	8	10980.0	11	18225.0	6	19440.0	5	6435.0	0	990.0
9	3	10485.0	16	17010.0	10	18720.0	1	6165.0	0	990.0
10	4	20340.0	9	31770.0	11	35550.0	3	11970.0	0	1980.0
11	9	18990.0	6	30420.0	7	33210.0	4	11160.0	1	1890.0
12	2	16290.0	8	25020.0	8	27270.0	0	9360.0	1	1260.0
13	3	12420.0	4	18540.0	6	19350.0	0	7110.0	1	540.0
14	2	8370.0	3	13140.0	5	13590.0	1	5040.0	0	360.0
15	1	4680.0	1	7380.0	1	8640.0	0	2970.0	0	360.0
16	0	1440.0	1	2160.0	0	3060.0	2	990.0	0	180.0
R = 2 (living, related)										
1	0	140.0	2	329.0	14	2821.0	4	1134.0	1	1144.5
2	1	156.0	0	368.0	7	3140.0	3	1268.0	2	1296.0
3	1	111.0	1	273.0	3	2325.0	5	927.0	1	963.0
4	0	162.0	2	396.0	4	3456.0	7	1336.0	1	1435.5
5	3	495.0	5	1215.0	31	10995.0	8	4230.0	5	4695.0
6	1	435.0	3	1095.0	21	10215.0	1	4095.0	1	4605.0
7	2	1170.0	2	3060.0	27	28485.0	8	11880.0	4	13590.0
8	0	1080.0	0	2970.0	14	26640.0	3	11385.0	1	13365.0
9	0	1080.0	1	2925.0	11	25515.0	1	11205.0	2	13230.0
10	0	2160.0	2	5580.0	9	49230.0	0	22320.0	0	26280.0
11	0	2160.0	0	5400.0	10	46800.0	1	22050.0	2	25740.0
12	0	1980.0	0	5040.0	7	39780.0	2	20070.0	1	23040.0
13	0	1800.0	1	3870.0	3	30060.0	1	16740.0	1	17820.0
14	0	1350.0	0	2790.0	2	21510.0	1	12690.0	1	12240.0
15	0	810.0	0	1890.0	0	13770.0	0	8100.0	1	7560.0
16	0	360.0	0	630.0	0	5130.0	0	2970.0	0	2700.0

[a] Source of data: Table 11.3.

Table 11.5

Fit of Models for the Graft-Survival Data With Collapsed Categories

Model	Degrees of freedom	Likelihood ratio	Probability
1. [RSM], [RO], [SO], [MO]	22	35.3	0.036
2. [RSM], [RMO], [SO]	20	28.4	0.100
Improvement in fit 2 versus 1	2	6.9	0.035
3. [RSM], [SMO], [RO]	14	23.5	0.053
4. [RSM], [RSO], [MO]	18	30.3	0.035
5. [RSM], [RSO], [SMO]	12	13.3	
6. [RSM], [RMO], [RSO]	16	23.5	

Based on these relative odds, one might consider collapsing the data with a view to capturing the essential patterns in terms of fewer parameters. Laird and Olivier decided to combine match grades 0, 1, and 2, and collapse the time dimension into five intervals: 0–1 month, 1–3 months, 3–6 months, 6 months to one year, and greater than one year. Table 11.5 shows the goodness of fit of a number of models to the collapsed data thus created. The difference between the likelihood ratio statistics of the first and second models, namely, 35.3 − 28.4 = 6.9, can be treated as a chi square with 22 − 20 = 2 degrees of freedom for purposes of testing whether the model with the lower degrees of freedom gives a significantly better fit to the data compared to the other model. Since this difference is relatively large (compared to the 5 percent level of the chi-square distribution with 2 degrees of freedom) we infer that the more inclusive model, namely, [RSM, RMO, SO], is preferable to the less inclusive one, [RSM, MO, RO, SO], on account of the significant improvement in the goodness of fit. This and other similar comparisons based on the figures presented in Table 11.5 lead one to choose [RSM, RMO, SO] as the best fitting parsimonious model. The relative odds reflecting the impact of the various factors on O can be computed from the expected values in the cells in accordance with the model [RSM, RMO, SO] and interpreted in the fashion commonly followed in interpreting results from experiments (see Problem 6).

3 Application of the General Linear Model Approach

Koch et al. (1972) show how to analyze survival rates using the method of weighted least squares. Gehan and Siddiqui (1973) demonstrate that the least-squares method is quite efficient for the analysis of survival data if the underlying hazard function or some transformation of it can be expressed as a linear function of convenient mathematical transformations of the parameters of interest. Examples of such situations include exponential sampling with parameter α, under which the hazard function $h(t)$ is a constant equal to α, and sampling from the Gompertz, under which the hazard function is $\alpha + \beta t$, where α and β are parameters of the Gompertz distribution. To introduce the Koch–Johnson–Tolley approach, consider the hypothetical data setup shown in Table 11.6. Imagine, for the moment, that r subpopulations were formed by combining categories of patient characteristics, method of treatment given, and the like. Within each subpopulation, each patient (subject) provides a failure time or a censored time. Using the information, it is possible to construct a life table for each subpopulation, showing for each time interval the number surviving to the beginning of, or entering, the interval, the number failing, and the number censored in the interval, and the number surviving to the beginning of the next. Thus, in Table 11.6, for group (subpopulation) i, n_{ij+} individuals entered the interval $[j - 1, j)$, of whom, n_{ij1} survived to the beginning of the next interval, n_{ij2} died, and n_{ij3} were censored during the time interval. Now let us assume that the triplets (number survived, number failed, and number censored) form a multinomial sample in each time interval for each group (subpopulation). Let us further assume that, within each group (subpopulation), each multinomial is uncorrelated with each other and that all multinomials within each group are independent of those in each other group. Next, let us imagine constructing response function scores for each subpopulation. The two-year survival rate or its logarithm are just two examples of such response functions. All such scores computed from the data are obviously mathematical functions of the observed n_{ijk}'s. Given the structure of the response function scores, it is possible to estimate (from the data) their expected values, variances, and covariances, using if necessary the delta method. These together with the factor combinations defining the subpopulations and the corresponding response function scores can be used for fitting a regression model.

Example 1. Koch *et al.* illustrate this approach using five-year survival rates for women under 65 years of age who were diagnosed as having breast cancer (the data were analyzed previously by Cutler and Myers, 1967).

Table 11.6

Layout of Data for the Application of the General Linear Model Approach

Subpopulation	Time interval	Number surviving	Number dying	Number withdrawn	Number alive at the beginning
i	$[j - 1, j)$	n_{ij1}	n_{ij2}	n_{ij3}	n_{ij+}

Subpopulations were formed by combining categories of the following three factors: node status (N) [with categories, clinically negative (0), and palpable (1)]; tumor size (T) [with categories, less than or equal to 2 cm (1), more than 2 cm but less than or equal to 4 cm (2), and more than 4 cm (3)]; and degree of skin fixation (S) [with categories, none (0), incomplete (1), and complete (2).] The estimated response function scores (five-year survival rates), their estimated standard errors, and the sample sizes are shown in Table 11.7 for the different subpopulations. Koch *et al.* fitted a number of models to the data in Table 11.7. One of their models has the "design" matrix X_1 shown in Fig. 11.1. In Table 11.8 are shown

$$(\text{estimate}/\text{estimated standard error})^2$$

of the different regression coefficients in the linear model specified in terms of the design matrix X_1. [These were computed for the present purpose using the SAS procedure for fitting general linear models (see SAS Institute, 1982). Koch

Table 11.7

Five-year Survival Rates and Their Standard Errors for 18 Groups of
Breast-Cancer Patients[a]

Subpopulation defined in terms of variables			Five-year survival rate	Standard error	Number in the sample
N	S	T			
0	0	1	0.88	0.024	195
0	0	2	0.77	0.028	226
0	0	3	0.62	0.050	96
0	1	1	0.78	0.049	72
0	1	2	0.67	0.050	89
0	1	3	0.49	0.069	53
0	2	1	0.95	0.034	41
0	2	2	0.74	0.042	114
0	2	3	0.51	0.057	78
1	0	1	0.63	0.099	24
1	0	2	0.58	0.066	55
1	0	3	0.57	0.065	59
1	1	1	0.93	0.069	15
1	1	2	0.67	0.086	30
1	1	3	0.38	0.095	26
1	2	1	0.71	0.171	7
1	2	2	0.47	0.129	15
1	2	3	0.39	0.079	38

[a] Source: Koch *et al.* (1972).

$$\begin{bmatrix}
1 & 1 & -1 & 1 & 0 & 0 & 1 & -1 & 0 & 0 & 0 & 0 & 0 & 0 & 0 & 0 & 0 \\
1 & 1 & -1 & 1 & 0 & 0 & 0 & 2 & 0 & 0 & 0 & 0 & 0 & 0 & 0 & 0 & 0 \\
1 & 1 & -1 & 1 & 0 & 0 & -1 & -1 & 0 & 0 & 0 & 0 & 0 & 0 & 0 & 0 & 0 \\
1 & 1 & -1 & -1 & 0 & 0 & 0 & 0 & 1 & -1 & 0 & 0 & 0 & 0 & 0 & 0 & 0 \\
1 & 1 & -1 & -1 & 0 & 0 & 0 & 0 & 0 & 2 & 0 & 0 & 0 & 0 & 0 & 0 & 0 \\
1 & 1 & -1 & -1 & 0 & 0 & 0 & 0 & -1 & -1 & 0 & 0 & 0 & 0 & 0 & 0 & 0 \\
1 & 0 & 2 & 0 & 1 & 0 & 0 & 0 & 0 & 1 & -1 & 0 & 0 & 0 & 0 & 0 & 0 \\
1 & 0 & 2 & 0 & 1 & 0 & 0 & 0 & 0 & 0 & 2 & 0 & 0 & 0 & 0 & 0 & 0 \\
1 & 0 & 2 & 0 & 1 & 0 & 0 & 0 & 0 & -1 & -1 & 0 & 0 & 0 & 0 & 0 & 0 \\
1 & 0 & 2 & 0 & -1 & 0 & 0 & 0 & 0 & 0 & 0 & 1 & -1 & 0 & 0 & 0 & 0 \\
1 & 0 & 2 & 0 & -1 & 0 & 0 & 0 & 0 & 0 & 0 & 0 & 2 & 0 & 0 & 0 & 0 \\
1 & 0 & 2 & 0 & -1 & 0 & 0 & 0 & 0 & 0 & 0 & -1 & -1 & 0 & 0 & 0 & 0 \\
1 & -1 & -1 & 0 & 0 & 1 & 0 & 0 & 0 & 0 & 0 & 0 & 0 & 1 & -1 & 0 & 0 \\
1 & -1 & -1 & 0 & 0 & 1 & 0 & 0 & 0 & 0 & 0 & 0 & 0 & 0 & 2 & 0 & 0 \\
1 & -1 & -1 & 0 & 0 & 1 & 0 & 0 & 0 & 0 & 0 & 0 & 0 & -1 & -1 & 0 & 0 \\
1 & -1 & -1 & 0 & 0 & -1 & 0 & 0 & 0 & 0 & 0 & 0 & 0 & 0 & 0 & 1 & -1 \\
1 & -1 & -1 & 0 & 0 & -1 & 0 & 0 & 0 & 0 & 0 & 0 & 0 & 0 & 0 & 0 & 2 \\
1 & -1 & -1 & 0 & 0 & -1 & 0 & 0 & 0 & 0 & 0 & 0 & 0 & 0 & 0 & -1 & -1
\end{bmatrix}$$

Fig. 11.1. Design matrix (referred to as X_1) used by Koch et al. (1972) in an analysis of the data in Table 11.9. The rows correspond to the rows of Table 11.7.

Table 11.8

Square of the Ratio of Effect Estimate to Its Standard Error: Results from Weighted Least-Squares Technique Applied to the Data in Table 11.7 Using the Design Matrix Shown in Fig. 11.1

Effect	Source (column of Fig. 11.1)	(Estimate/standard error)^2
Skin fixation		
Linear effect	2	5.02[a]
Nonlinear effect	3	0.22
Node status		
first category of skin fixation	4	8.10[b]
second category of skin fixation	5	7.19[b]
third category of skin fixation	6	2.29
Tumor size		
Linear effect $S = 0, N = 0$	7	22.26[b]
Nonlinear effect $S = 0, N = 0$	8	0.26
Linear effect $S = 0, N = 1$	9	11.77[b]
Nonlinear effect $S = 0, N = 1$	10	0.29
Linear effect $S = 1, N = 0$	11	43.88[b]
Nonlinear effect $S = 1, N = 0$	12	0.04
Linear effect $S = 1, N = 1$	13	0.26
Nonlinear effect $S = 1, N = 1$	14	0.05
Linear effect $S = 2, N = 0$	15	21.87[b]
Nonlinear effect $S = 2, N = 0$	16	0.02
Linear effect $S = 2, N = 1$	17	2.89
Nonlinear effect $S = 2, N = 2$	18	0.25

[a] Significant effect at 5% level.
[b] Significant effect at 1% level.

et al. used a special program in their analysis. The FUNCAT procedure of SAS can also be used instead.] The estimated node status effects conditional on S are: 0.055, when $S = 0$, 0.077, when $S = 1$, and 0.068, when $S = 2$. Since these figures are approximately of the same magnitude, it is reasonable to consider averaging them rather than focusing on each conditional effect separately. Note that to replace in the model the three conditional effects of node status by their average, one simply merges the columns of the "design" matrix corresponding to the conditional effects. Similarly, an examination of the linear component of the effect of tumor size, conditional on S and N,

$S = 0, N = 0$	$S = 0, N = 1$	$S = 1, N = 0$	$S = 1, N = 1$	$S = 2, N = 0$	$S = 2, N = 1$
0.130	0.145	0.220	0.030	0.275	0.160

suggests that it might be reasonable to consider replacing them by the following two average effects: one averaged over $S = 0, N = 0$; $S = 0, N = 1$; and $S = 2, N = 1$, and the other averaged over $S = 2, N = 0$, and $S = 1$, $N = 0$.

Also the figures in Table 11.8 clearly suggest that we may discard the nonlinear components of the tumor size and the skin-fixation effects. Figure 11.2 shows a design matrix $\mathbf{X}_2$ that incorporates these suggestions. The results of fitting the model specified in terms of the design matrix in Fig. 11.2 are shown in Table 11.9. The residual sum of squares shown in

$$\begin{bmatrix}
1 & 1 & 1 & 1 & 0 \\
1 & 1 & 1 & 0 & 0 \\
1 & 1 & 1 & -1 & 0 \\
1 & 1 & -1 & 1 & 0 \\
1 & 1 & -1 & 0 & 0 \\
1 & 1 & -1 & -1 & 0 \\
1 & 0 & 1 & 0 & 1 \\
1 & 0 & 1 & 0 & 0 \\
1 & 0 & 1 & 0 & -1 \\
1 & 0 & -1 & 0 & 0 \\
1 & 0 & -1 & 0 & 0 \\
1 & 0 & -1 & 0 & 0 \\
1 & -1 & 1 & 0 & 1 \\
1 & -1 & 1 & 0 & 0 \\
1 & -1 & 1 & 0 & -1 \\
1 & -1 & -1 & 1 & 0 \\
1 & -1 & -1 & 0 & 0 \\
1 & -1 & -1 & -1 & 0
\end{bmatrix}$$

Fig. 11.2. Design matrix referred to as $\mathbf{X}_2$ in text.

Table 11.9

Analysis by the Weighted Least-Square Method of the Data in Table 11.7 Using the Design Matrix Given in Table 11.2

Effect	Number of degrees of freedom	Sum of squares
Node status	1	20.48
Skin fixation	1	6.37
Tumor size		
Average linear effect in (0, 0), (0, 1), and (2, 1) combinations of (S, N)	1	39.31
Average linear effect in (1, 0) and (2, 0) combinations of (S, N)	1	75.25
Residual	13	2.76

Table 11.9 can be treated as a goodness-of-fit chi square. In the present case, the small magnitude of this statistic suggests that the corresponding model adequately accounts for the patterns in the data. [Koch *et al.*, (1972) recommend that the goodness-of-fit chi square for the excluded degrees of freedom (equal to 13 in the example) should not exceed 3.84 (the 5 percent point of the chi square with 1 degree of freedom to ensure that the residual does not contain any hidden, but individually significant, component. Uncritical adoption of this recommendation may result in "over fitting" the data (Namboodiri and West, 1978).]

Example 2. The data for this example come from a large study (Kahn, 1966) of the relationship between smoking and death rates. About 248,000 male policy holders insured by the government answered mailed questionnaires about their smoking habits. The data are shown in Table 11.10 for men of age 35–44, 45–54, 55–64, and 65–74 who reportedly smoked on average 0, 1–9, 10–20, 21–39, or over 39 cigarettes per day. The upper panel of the table shows, for each subpopulation thus formed, the number of deaths, and the lower panel shows 1000 times the corresponding probabilities of dying, computed in accordance with the usual life table procedure. A cursory look at the data in the lower panel of the table shows that the probability of dying rises sharply with each additional number of cigarettes smoked. Also, as we would expect, these probabilities show a sharply increasing trend with age within each smoking category. A hypothesis of interest may be that the smoking effect is the same in each age class, i.e., there is no interaction effect involving age and smoking. The following model reflects this hypothesis:

$$p_{ij} = \alpha_i \beta_j \tag{11.29}$$

Table 11.10

Number of Deaths and Annual Probabilities of Dying ($\times 10^3$): U. S. Veterans[a]

Age (years)	Reported number of cigarettes smoked per day				
	None	1–9	10–20	21–39	over 39
	Number of deaths				
35–44	47	7	90	83	10
45–54	38	11	67	80	14
55–64	2617	389	2117	1656	406
65–74	3728	586	2458	1416	256
	Annual probabilities of dying ($\times 1000$)				
35–44	1.27	1.63	1.99	2.66	3.26
45–54	2.64	6.23	6.64	8.91	11.60
55–64	10.56	14.35	18.50	20.87	27.40
65–74	24.11	35.76	42.26	49.40	55.91

[a] Source: Kahn (1966).

where p_{ij} is the (annual) probability of dying for those in the ith age group and the jth smoking class, $\alpha_1, \alpha_2, \ldots$ are the age effects, and $\beta_1, \beta_2, \ldots$ are the smoking effects. Taking logarithms we obtain

$$\ln p_{ij} = \ln \alpha_i + \ln \beta_j \qquad (11.30)$$

Let us assume that the p's (omitting the multiplier 1000) are independent binomials. [Within an individual subpopulation, the probability of dying is likely to vary over individuals. If such a variation exists, it would make the variance of p less than the binomial variance. But with small p's (in the present data between 0.00127 and 0.05591), the difference between the actual variance and the binomial variance is likely to be negligible.] Assuming that the p's are binomial, the weight for each $\ln p$ is the observed number of deaths, the estimated variance of $\ln p_{ij}$ being approximately $1/(t_{ij+}p)$ for fixed total exposure t_{ij+}.

Fitting the model (in obvious notation) $y_{ij} = a_i + b_j + e_{ij}$ to the data in Table 11.10, using the weighted regression procedure of SAS, yields the following goodness-of-fit statistic:

$$\sum w(y - \hat{y})^2 = 13.93217 \qquad (11.31)$$

which may be treated as a chi square with 12 degrees of freedom. The magnitude of this figure indicates no lack of fit. We may therefore infer that the smoking effect remains age-invariant for males between 35 and 75. If we subtract the estimate of b_0 from that of b_j and exponentiate the difference we get an estimate of the multiplier effect of the jth class of smoking on the probability of dying. The multipliers thus estimated for the different smoking classes are shown below: 1.436 for 1–9 cigarettes per day, 1.756 for 10–20 cigarettes per day, 2.026 for 21–39 cigarettes per day, and 2.504 for over 39 cigarettes per day.

Problems and Complements

1. The notion that survival time may be dependent on one or more covariates (factors) is ubiquitous. Thus, it is common to assume that patient survival is dependent on the type and strength of medication given. As another example, in an industrial life-testing experiment, the length of life of the device being tested is a function of the stress conditions that prevail. Feigle and Zelen (1965) proposed for exponential samples linear and log linear specifications of the dependence of survival time on an explanatory variable (covariate, factor). The linear model can be expressed as $E(T) = \alpha + \beta z$, where z is the explanatory variable. The corresponding log-linear model is $E(T) = \alpha \exp(\beta z)$ or, equivalently, $\ln E(T) = \ln \alpha + \beta z$. A problem with the linear model is that there is no guarantee against getting a negative estimate for $E(T)$. The log linear specification does not have this problem. A discussion of both of these models when there is no censoring is available in Feigle and Zelen (1965). The censored case of the linear model has been discussed by Zippin and Armitage (1966), among others, while the corresponding log linear setup has been discussed by Glasser (1967).

2. Miller (1981) refers to models with surviving fractions. If p is the probability of dying, then $1 - p$ is called the *surviving fraction*. Models with surviving fractions are sometimes used for short-term experiments where one does not hypothesize that the survival function necessarily approaches zero. One may specify that the survival function levels off at $1 - p$ instead of decreasing to zero. Thus, in the study of first birth intervals, one may wish to stipulate that a certain fraction of women will be childless throughout the reproductive period. Under exponential sampling, the contribution to the likelihood from an uncensored case is $p\alpha e^{-\alpha t}$, whereas the corresponding contribution from a censored case is $(1 - p) + pe^{-\alpha t}$. For estimation one uses in a straightforward fashion maximum likelihood methods.

3. Suppose the probability of dying is dependent on an explanatory variable x in accordance with the logistic pattern

$$p(x) = p(\text{dying} \,|\, x) = \frac{\exp(\alpha + \beta x)}{1 + \exp(\alpha + \beta x)}$$

Then under exponential sampling, the contribution from an uncensored case to the likelihood is $p(x)\theta \exp(-\theta t)$ whereas that from a censored case is $1 - p(x) + p(x) \exp(-\theta t)$ (Miller, 1981). Maximum likelihood methods can be applied for estimation.

4. The dependence of mortality on one or more factors can be modeled in terms of a corresponding dependence of the force of mortality on the factors. Suppose the age (duration) dimension has been categorized into, say, I segments, and mother's education, say, has been defined in terms of, say, J categories. Then a simple model for the force of mortality associated with children who survive until the beginning of the age group $[a_{i-1}, a_i)$ is

$$\mu_{ijk} = \exp(\beta + \beta_i^A + \beta_j^B)$$

Table 11.11

Reanalysis of the Laird–Olivier Data Using the Weighted
Least-Squares Method

Model (factors included)	Number of degrees of freedom	$\Sigma\, W(y - \hat{y})^2$
1. $S, M, R,$	21	36.44
2. $S, M, R, S \times R$	17	32.50
3. $S, M, R, S \times M$	13	21.92
4. $S, M, R, R \times M$	19	23.61
5. $S, M, R, S \times R, R \times M$	9	16.20
6. $S, M, R, R \times M, S \times M$	11	10.50
7. $S, M, R, R \times M, S \times R$	15	20.58

or

$$\ln \mu_{ijk} = \beta + \beta_i^A + \beta_j^B$$

where β is common for all children (in all ages and all categories of mother's education), β_i^A is common for all children of age a_{i-1} and whose mothers are of education level j. Notice that in this simple model the impact of mother's education on the force of mortality is stipulated to be time invariant. A corresponding time-dependent formulation is

$$\ln \mu_{ijk} = \exp(\beta + \beta_i^A + \beta_j^B + \beta_{ij}^{AB})$$

in which the last term on the right-hand side is a parameter specific for the (i, j) cell of the $A \times B$ classification. A nonzero value for these parameter signifies that the impact of mother's education on the force of mortality varies with the child's age.

5. We have just seen an example of an explanatory factor whose impact on the force of mortality changes with survival time. The value of the factor itself does not change—it is its impact on the force of mortality that changes with survival time. To be contrasted with such explanatory factors are those that change their values over time. An example is health status of the breastfeeding mother. Incorporation of such factors in the model demands attention to issues that are not pertinent to other type of explanatory factors. Thus with respect to the health status of the mother, one needs to ask whether there are carryover effects, e.g., the health status of the mother at the time of birth affecting the force of mortality of the child beyond one year of age. Some help in thinking about such issues can be obtained from the literature on experimental designs using the same subjects repeatedly (see, e.g., Namboodiri, 1972).

6. Using the parameter estimates for the model $[RSM]$, $[RMO]$, $[SO]$ fitted to the collapsed graft-survival data referred to in Section 2, compute relative odds so as to portray the time effect and the effects of R and M on O.

7. Analyze the collapsed graft-survival data using the technique employed in the text for the smoking data and compare the results with those presented in the text. See Table 11.11.

Bibliographic Notes

According to Cox and Oakes (1984), the exponential distribution was probably studied first in connection with the kinetic theory of gases. Lawless (1982) points out that the exponential distribution was the survival time model for which statistical methods were first developed. Sukhatme (1937) and Epstein and Sobel (1953) are credited with early results that popularized the use of the exponential model in industrial life testing. For a good exposition of the idea of piecewise models to represent survival distributions, reference may be made to Elandt–Johnson and Johnson (1980). Laird and Olivier (1981) and Cox and Oakes (1984) show that log-linear models for the cell means of contingency tables with Poisson data are equivalent to log-linear hazard models for survival data, when the survival distribution is specified in the form of piecewise exponential and the covariates are all categorical. They also show that the likelihoods of piecewise exponential survival data and Poisson contingency data are equivalent. The way is thus cleared for the application of log-linear techniques to the analysis of survival data. The basic model used in the work of Laird and Olivier was proposed earlier by Holford (1976). The general linear model approach to the analysis of categorical data was developed by Grizzle *et al.* (1969). Koch *et al.* (1972) illustrate the application of the approach in the analysis of survival data.

An enormous literature is available on parametric representation of birth intervals in fertility analysis. Reference has already been made to the pioneering works of Potter (1963, 1967, 1969) and of Sheps (1965). An often-cited publication in this connection is that of Perrin and Sheps (1964) on human reproduction as a renewal process. Menken (1975) has provided an excellent nontechnical review of biometric models of varying complexity, with illustrations of their use in fertility research. A more complete treatment is available in Sheps and Menken (1973). Singh and his collaborators have examined the use of the exponential model in the analysis of birth intervals (see, e.g., Singh, 1968, and Singh *et al.*, 1974, 1979).

Chapter 12 | Proportional Hazards and Related Models

We have seen how to test whether different samples of failure times represent different survival functions. The method involves estimating survival functions for each sample and then making comparisons either directly or through summary statistics. Applying this procedure to answer questions such as whether the timing of contraceptive sterilization varies by the couple's demographic or socioeconomic characteristics involves subclassifying the sample on the basis of the covariates of interest and estimating survival functions in each subclass separately for comparison. When none of the subclasses is small this procedure is workable. But very often many of the subclasses are likely to be small, especially when several covariates are to be taken into account simultaneously. Under such circumstances it would be advisable to use for comparison of survival functions comprehensive models in which the effects of factors affecting failure times are represented by unknown parameters. In this chapter we review several such models. Typically we assume that, for each individual in the sample, information is available on a number of characteristics that affect failure time. The term covariate is often used for such characteristics. An alternative term is "explanatory variable." To give an example, Trussel and Hammerslough (1983) hypothesized that in Sri Lanka, the hazard (rate) of death at infant ages depended on mother's education, father's education, age of mother at birth (of the child), residence (rural, urban, estate), ethnicity of mother, type of toilet facility, source of water supply, time period of birth, and sex of the child. To give another example, Menken and her colleagues (1981) examined the dependence of the hazard of marriage dissolution in the United States on a number of sociodemographic factors [time period of marriage, age at marriage, education, religion, and type of first birth (illegitimate, premaritally conceived, other)]. Sometimes the dependence may be on a dichotomous variable, such as whether the subject is in the treatment or the control group. Not infrequently, a covariate may be time dependent. For example a treatment under study may not be applied until some time after the start of the study. In such a situation, it is common to give the covariate representing the treatment variable the value 0 until the treatment starts and 1 afterward. To give another example, in the study of infant mortality, mother's educational background is time invariant if all women complete their education before childbearing starts, but if the pursuit of education continues after childbearing commences, then the covariate becomes time varying or time dependent. We discuss the case of time-invariant covariates first.

1   Basic Ideas

To fix ideas let us imagine that we are studying infant mortality. Let $z_1, \ldots, z_k$ be k covariates under consideration. Clearly, the dependence of failure time (age at death) on these covariates can be expressed abstractly in a number of ways. For example, we may think of age at death or a transformation of it (e.g., its logarithm) as a function of the z's; if we add a residual to that function we have what resembles the usual regression model. Alternatively, we may express the hazard rate as a function of age and the covariates. Another option might be to consider a summary characteristic of the survival function (e.g., the proportion surviving to age a) as the dependent variable. This option was considered in Chapter 11. This chapter is devoted to the hazard function approach.

In the usual life (mortality) table, the hazard rate (force of mortality) is assumed to be a function of age only. But the fact that sometimes we construct one life table for males and another for females implies that the hazard rate is regarded as a function of age and sex. Similarly, when we recognize the need to construct separate life tables for subpopulations defined in terms of race and sex, we are admitting that the hazard rate is a function of age, sex, and race. In such situations, it is often convenient to think of developing models in two parts. First, we develop a model that corresponds to a reference group (e.g., black females). Then represent the change in the hazard rate induced by a shift in race (from black to white) and/or sex (from female to male). Stated more abstractly, we define the covariates (e.g., sex, race) in such a way that all of them take the value 0 for the reference group, and then represent, in parametric form, the change induced by a nonzero covariate.

If we consider the simplest case involving one dichotomous covariate z, taking value 0 for the reference group and 1 for the other, we may represent the hazard function for the reference group by $h_0(a)$ and that for the other (that is, when $z = 1$) by

$$h_1(a) = h_0(a) \exp(\beta) \qquad (12.1)$$

for example. If z represents two levels of a treatment, the above formulation implies that all individuals under treatment $z = 0$ have a common hazard function $h_0(a)$, which would be multiplied by a factor $\exp(\beta)$ if they were to be put under treatment $z = 1$. In this formulation the covariate acts on the hazard rate multiplicatively. Interpretation of the same in terms of survival time is not difficult if we remember that increasing hazard rate corresponds to decreasing survival time. Also, if $S_0(a)$ is the baseline survival function corresponding to the hazard function $h_0(t)$, then the survival function corresponding to $z = 1$ is

$$S_1(a) = [S_0(a)]^{\exp(\beta)} \qquad (12.2)$$

This model is flexible enough for many applications. Notice that in this formulation the baseline hazard function is viewed as an arbitrary function of age (time). [Of course, if the data on hand can be represented by a specific hazard function such as that of any known distribution, e.g., log normal, then one replaces the arbitrary $h_0(a)$ by the corresponding specific hazard function.] The model described above belongs to a family called *proportional hazards* family, which has the property that different individuals (e.g., one with $z = 0$ and one with $z = 1$) have hazard functions proportional to one another. Extending to the case of two binary covariates, it is easily seen that there are four proportional hazard functions involved:

$$h_0(a); \qquad h_0(a) \exp(\beta_1 + \beta_2); \qquad h_0(a) \exp(\beta_1); \qquad h_0(a) \exp(\beta_2)$$

Of these, the first one is the baseline hazard function, the second one corresponds to individuals with both covariates taking the value 1, and so on. In general, a proportional hazards model can be expressed in the form

$$h(a \,|\, z_1, \ldots, z_k) = h_0(a)\psi(z_1, \ldots, z_k) \qquad (12.3)$$

where the multiplier $\psi(z_1, \ldots, z_k)$ does not vary with age (time) and is positive. A particular subfamily of this is the following:

$$h(a \,|\, z_1, \ldots, z_k) = h_0(a)\exp(\sum \beta_j z_j) \qquad (12.4)$$

where the β's are unknown regression coefficients. Notice that $\exp(\sum \beta_j z_j)$ is always positive. This is an advantage, for it permits the specification of the model without any further restrictions on the parameters. In other formulations, such as a linear function of parameters, additional restrictions that the functions be positive must be incorporated. Two basic assumptions are involved in proportional hazards models:

1. All individuals with a given configuration of values for the covariates $z_1, \ldots, z_k$ have identical hazards functions.

2. The hazards functions of any two individuals differing in the configuration of values of the covariates have parallel age (time) patterns.

The former can often be met by appropriate choice of covariates. The second assumption, which, incidentally, is known as the *proportionality assumption* cannot be satisfied that easily. Formal tests of the proportionality assumptions are discussed elsewhere in the book.

2   Estimation and Testing Hypotheses

We now consider inference about the multipliers in

$$h(t \mid z_1, \ldots, z_k) = h_0(t)\psi(z_1, \ldots, z_k) \tag{12.5}$$

To fix ideas, let us consider a simple case involving no censoring. Let us assume that survival times have continuous distributions and are recorded exactly, so that ties are unlikely.

Suppose the sample size is four (purposely kept small to facilitate exposition). Let the individuals be named A, B, C, and D, and let their failure times be t_A, t_B, t_C, and t_D, respectively. Suppose that $t_C < t_D < t_A < t_B$. For convenience let us write t_1 for t_C, t_2 for t_D, t_3 for t_A, and t_4 for t_B. Then the ordered failure times are

$$t_1 < t_2 < t_3 < t_4$$

Clearly, just before the occurrence of the first failure, all four individuals were at risk. Put differently, the risk set at t_1 was

$$R_1 = R(t_1) = \{A, B, C, D\}.$$

Similarly, the risk sets just before t_2, t_3, and t_4, were, respectively,

$$R_2 = R(t_2) = \{A, B, D\},$$
$$R_3 = R(t_3) = \{A, B\},$$

and

$$R_4 = R(t_4) = \{B\}.$$

Likelihood

In the present case, the likelihood on the data is a constant multiple of the joint probability of C failing at t_1, followed by D, A, and B in that order at t_2, t_3, and t_4, respectively. It is easy to see that just before t_1, we would put the probability of C failing at t_1, given the risk set $R(t_1)$, as

$$\frac{h_C(t_1)}{h_A(t_1) + h_B(t_1) + h_C(t_1) + h_D(t_1)} \tag{12.6}$$

Similarly, we have, given the respective risk sets, for the probability of D failing at t_2, A at t_3, and B at t_4:

$$h_D(t_2)/[h_A(t_2) + h_B(t_2) + h_D(t_2)] \tag{12.7}$$

$$h_A(t_3)/[h_A(t_3) + h_B(t_3)] \tag{12.8}$$

and

$$h_B(t_4)/h_B(t_4) \tag{12.9}$$

respectively. Writing $\psi_1, \psi_2, \psi_3, \psi_4$ for the multipliers in (12.5) for C who failed first, for D who failed second, for A who failed third, and for B who failed last, we notice that (12.6), (12.7), (12.8), and (12.9) simplify, respectively, to

$$\psi_1/(\psi_1 + \psi_2 + \psi_3 + \psi_4) \tag{12.10}$$

$$\psi_2/(\psi_2 + \psi_3 + \psi_4) \tag{12.11}$$

$$\psi_3/(\psi_3 + \psi_4) \tag{12.12}$$

and

$$\psi_4/\psi_4 \tag{12.13}$$

We thus obtain the likelihood on the given data as the product of these four conditional probabilities. Therefore, the steps involved in writing the likelihood are

1. Order the failure times (which are all distinct, since we are discussing the case with no ties) from the shortest to the longest:

$$t_1 < t_2 < \cdots < t_k$$

2. At each failure time figure out the risk set:

$$R_1, \ldots, R_k$$

3. At each failure time calculate the ratio of the multiplier attached to the hazard rate of the individual who fails there to the sum of the corresponding multipliers for those in the relevant risk set:

$$\psi_i \bigg/ \sum_{i \in R_i} \psi_i \tag{12.14}$$

where the summation in the denominator is over the risk set at t_i.

4. Obtain the product of the ratios thus obtained. In the notation of (12.14) the product can be expressed as

$$\prod_i \left(\psi_i \bigg/ \sum_{i \in R_i} \psi_i \right) \tag{12.15}$$

Likelihood When There Is Censoring

Let us suppose that in the illustrative case examined above, individual D was censored after C failed but before A did. The unmeasured failure time of D could fall in reality between the failure times of C and A, between the failure times of A and B, or after B's failure time. To obtain the likelihood on the data we add the likelihoods of the three possibilities. A little algebra reveals that the likelihood thus obtained is equal to

$$\frac{\psi_C}{\psi_A + \psi_B + \psi_C + \psi_D} \frac{\psi_A}{\psi_A + \psi_B} \frac{\psi_B}{\psi_B}$$

The steps involved in the construction of likelihood when there is censoring are thus essentially the same as those described above for the situation involving no censoring, except that the numerators of the ratios involved pertain exclusively to failure times; censored cases enter the picture, however, through the risk sets.

Likelihood When There Are Ties

If several failures are tied at a particular time point, a recommended approximation is to incorporate in the likelihood a factor (12.14) from each case in the tied set. The denominator of each contribution would be the same; the numerator would differ depending upon the variation of the cases with respect to the covariate values.

Example. Suppose the following data came from an experiment in which subjects were assigned at random to a treatment group and control:

Treatment group	1	3	3*	4	5	5	9*
Control group	2	3*	4	6	6		

where * indicates censoring. The steps one might use in constructing the likelihood on the data are presented in Table 12.1. We define a covariate that takes the value 1 for those in the treatment group and 0 for those in the control group. We further assume that the hazard rate is the product of an arbitrary function $h_0(t)$, where t is the duration since randomization and $\psi(z)$ a function of z. To be more specific we use $e^{\beta z}$ for $\psi(z)$, where β is an unknown parameter. Note that under the specific setup, $\psi(0) = 1$, and $\psi(1) = e^{\beta}$.

Column 1 of Table 12.1 shows the observations (in the pooled sample) arranged in ascending order (with the smallest at the top). Columns 2 and 3 repeat the entries in column 1, with column 2 pertaining to the observations in

Table 12.1

Illustration Showing the Construction of the Likelihood Function under the Proportional Hazards Model

Ordered observations			Multiplier		Cumulative from below of sum of columns 4 and 5	Contribution to likelihood = ratio of column 4 or column 5 to column 6
Pooled sample (1)	Treatment (2)	Control (3)	Treatment (4)	Control (5)	(6)	(7)
1	1		e^β		$5 + 7e^\beta$	$e^\beta/(5 + 7e^\beta)$
2		2		1	$5 + 6e^\beta$	$1/(5 + 6e^\beta)$
3	3		e^β		$4 + 6e^\beta$	$e^\beta/(4 + 6e^\beta)$
3*	3*	3*	$\{e^\beta\}$	$\{1\}$	$4 + 5e^\beta$	
4	4	4	e^β	1	$3 + 4e^\beta$	$e^\beta/(3 + 4e^\beta); 1/(3 + 4e^\beta)$
5, 5	5, 5		$e^\beta; e^\beta$		$2 + 3e^\beta$	$e^\beta/(2 + 3e^\beta); e^\beta/(2 + 3e^\beta)$
6, 6		6, 6		1, 1	$2 + e^\beta$	$1/(2 + e^\beta); 1/(2 + e^\beta)$
9*	9*		$\{e^\beta\}$		e^β	

the treatment group, and column 3 to those in the control group. Column 4 (column 5) shows the multiplier $\psi(z)$ associated with the entries in column 2 (column 3). All multipliers associated with censored observations are shown in braces. Column 6 shows the cumulatives, from below, of the *sum* of the entries in columns 4 and 5. Column 7 shows the ratio of the entry in column 4 or column 5 as the case may be to the corresponding entry in column 6, with no entry shown for censored observations. The likelihood is the product of the entries in column 7.

Several remarks are in order regarding the procedure for likelihood construction illustrated in Table 12.1. First, it should be remarked that the procedure does not use the observed actual values of the failure or censoring times; it uses only the rank order of the observations. Second, the procedure is based on the assumption that the failure times are continuous and that the ties present are due to crudity of measurement. Kalbfleisch and Prentice (1973) show that the likelihood function (12.15) can be derived as a marginal likelihood function based on the probability distribution of rank statistic for the data when there is no censoring and no ties. The approximation used in the illustration of Table 12.1 was first suggested by Peto (1972). If items A and B are observed to fail at τ, given the risk set $\{A, B, C, D\}$, the combined contribution to the likelihood from the observations at τ is

$$\frac{\psi_A}{\psi_A + \psi_B + \psi_C + \psi_D} \frac{\psi_B}{\psi_B + \psi_C + \psi_D} + \frac{\psi_B}{\psi_A + \psi_B + \psi_C + \psi_D} \frac{\psi_A}{\psi_A + \psi_C + \psi_D}$$

The Peto approximation mentioned above is obtained by letting all sums in the denominators include all items in the risk set, thus yielding

$$2\psi_A\psi_B/[\psi_A + \psi_B + \psi_C + \psi_D]^2$$

By generalizing from this to the case of d deaths tied at τ, given that the risk set at τ is $R(\tau)$ and on using the labels $1, 2, \ldots, d$ for the deaths, we find

$$d!\,\psi_1\psi_2 \cdots \psi_d \bigg/ \left[\sum_{i \in R(\tau)} \psi_i\right]^d \tag{12.16}$$

This approximation is considered satisfactory, except when ties are heavy.

Breslow's (1974) Approximation

Breslow (1974) has suggested a simpler approach to likelihood construction when there are ties. He starts with the specification of the hazard function in the same fashion as described above, namely, a baseline hazard function multiplied by a function of covariates. He then restricts the baseline hazard function to be a step function with discontinuities at each *observed*, distinct failure time. For censored times in $[\tau_{i-1}, \tau)$ it is assumed that nothing is known beyond the beginning of the interval, or, equivalently, that all censorings occur at the beginning of the interval. To illustrate this procedure, suppose that starting at time 0 four items are put on test, that one of them, say, A, fails at t_1, two, say, B and C, fail at t_2, and one, say, D, is censored between t_1 and t_2. Let the baseline hazards be

$$h_0(t) = \theta_i \qquad \text{for} \quad t_i < t \leq t_{i+1}$$

where $t_0 = 0 < t_1 < t_2 < t_3 = \infty$ are distinct failure times. The contributions from A, B, C, and D to the likelihood are respectively

$$\theta_1\psi_A \exp[-(t_1 - t_0)\theta_1\psi_A]$$

$$\theta_2\psi_B \exp[-(t_1 - t_0)\theta_1\psi_B - (t_2 - t_1)\theta_2\psi_B]$$

$$\theta_2\psi_C \exp[-(t_1 - t_0)\theta_1\psi_C - (t_2 - t_1)\theta_2\psi_C]$$

and

$$\exp[-(t_1 - t_0)\theta_1\psi_D].$$

Taking the logarithm of the product of these contributions, we obtain the log likelihood on the data as

$$L = \ln \theta_1 + 2 \ln \theta_2 + \ln(\psi_A\psi_B\psi_C) - [(t_1 - t_0)\theta_1(\psi_A + \psi_B + \psi_C + \psi_D)]$$
$$- (t_2 - t_1)\theta_2(\psi_B + \psi_C)$$

For fixed ψ's, L is maximum when

$$\theta_1 = 1/(t_1 - t_0)(\psi_A + \psi_B + \psi_C + \psi_D)$$

and

$$\theta_2 = 2/(t_2 - t_1)(\psi_A + \psi_C)$$

Substitution yields the log likelihood maximized over the ψ's as

$$L = -\ln(\psi_A + \psi_B + \psi_C + \psi_D) - 2\ln(\psi_B + \psi_C) + \ln\psi_A + \ln\psi_B + \ln\psi_C$$

$$(12.17)$$

except for an additive constant. It is easily seen that the corresponding likelihood is proportional to

$$\frac{\psi_A}{\psi_A + \psi_B + \psi_C + \psi_D} \frac{2\psi_B\psi_C}{(\psi_B + \psi_C)^2} \qquad (12.18)$$

which in turn is the same as the Peto approximation to the likelihood mentioned above.

Generalizing (12.17) we obtain

$$L = \sum_u \ln\psi_i - \sum_{i=1}^{k} d_i \ln \sum_{j\in R_i} \psi_j \qquad (12.19)$$

for distinct failure times $t_0 = 0 < t_1 < \cdots < t_k < t_{k+1} = \infty$, with d_i failures at t_i, where the risk set is R_i, Σ_u denoting summation over the uncensored observations. In the log-linear setup, i.e., when $\ln\psi$ is linear in a set of parameters, $\Sigma_u \ln\psi_i$ simplifies considerably. This is because when $\psi = \exp(\Sigma_g \beta_g z_g)$, $\ln\psi = \Sigma_g \beta_g z_g$, so that

$$\sum_i \ln\psi_i = \sum_i \left[\sum_g \beta_g z_{gi}\right] = \sum_g \beta_g \left(\sum_i z_{gi}\right)$$

To illustrate the point let us consider the setup in Table 12.1. Taking the logarithm of the product of the entries in column 7 we obtain the log likelihood as

$$L = 5\beta - [\ln(5 + 7e^\beta) + \ln(5 + 6e^\beta) + \ln(4 + 6e^\beta)$$
$$+ 2\ln(3 + 4e^\beta) + 2\ln(2 + 3e^\beta) + 2\ln(2 + e^\beta)]$$

which can be expressed as

$$L = \left(\sum_i m_{i1}\right)\beta - \sum_i [m_{i+}\ln(R_{i0} + R_{i1}e^\beta)] \qquad (12.20)$$

where m_{i1} is the number of observed failures at t_i in the sample with $z = 1$, m_{i0} the corresponding number in the sample with $z = 0$, $m_{i+} = m_{i1} + m_{i0}$

is the number of failures at t_i in the pooled sample, and R_{ij} is the size of the risk set (that is, the number of items in the risk set) at t_i in the sample with $z = j$.

Example. To provide a numerical example, we turn to data of Freireich et al. (1963), which were discussed by Cox in his 1972 paper. The data (shown below) concern the remission of leukemia patients in weeks after the treatment started. Patients were divided into two groups, one receiving treatment and the other serving as control. (The treatment involved administration of the drug 6-MP.) Some of the observations are right-censored, and are indicated by (*):

Treatment group ($z = 1$)	6, 6, 6, 6*, 7, 9*, 10, 10*, 11*, 13, 16, 17*, 19*, 20*, 22, 23, 25*, 32*, 32*, 34*, 35*
Control group ($z = 0$)	1, 1, 2, 2, 3, 4, 4, 5, 5, 8, 8, 8, 11, 11, 12, 12, 15, 17, 22, 23

Table 12.2 shows the values of m_{i0}, m_{i1}, m_{i+}, R_{i0}, and R_{i1}. To test the hypothesis that $\beta = 0$, we evaluate the the first and second derivatives of the log likelihood at $\beta = 0$, divide the former by the square root of the negative of the latter, and treat it as a normal deviate for testing purposes. It is easy to see from (12.20) that the first derivative of L is

$$U(\beta) = \sum_i m_{i1} - \sum_i [m_{i+} R_{i1} e^\beta / (R_{i0} + R_{i1} e^\beta)] \tag{12.21}$$

and the negative of the second derivative is

$$I(\beta) = \sum_i [m_{i+} R_{i0} R_{i1} e^\beta / (R_{i0} + R_{i1} e^\beta)^2] \tag{12.22}$$

The value of $U(\beta)$ when $\beta = 0$ can be expressed as

$$U(0) = \sum_i \left[m_{i1} - m_{i+} \frac{R_{i1}}{R_{i+}} \right] \tag{12.23}$$

and the value of $I(0)$ is seen to be

$$I(0) = \sum [m_{i+} R_{i0} R_{i1} / (R_{i0} + R_{i1})^2] \tag{12.24}$$

From Table 12.2, we have

$$\sum m_{i1} = 21$$

and

$$\sum \frac{m_{i+} R_{i1}}{R_{i0} + R_{i1}} = 10.79$$

Table 12.2

Analysis of the Data from Freirich et al. (1963)[a]

t_i	m_{i0}	m_{i1} [a]	m_{i+} [b]	R_{i0}	R_{i1}	R_{i+}
1	0	2	2	21	21	42
2	0	2	2	21	19	40
3	0	1	1	21	17	38
4	0	2	2	21	16	37
5	0	2	2	21	14	35
6	3	0	3	21	12	33
7	1	0	1	17	12	29
8	0	4	4	16	12	28
10	1	0	1	15	8	23
11	0	2	2	13	8	21
12	0	2	2	12	6	18
13	1	0	1	12	4	16
15	0	1	1	11	4	15
16	1	0	1	11	3	14
17	0	1	1	10	3	13
22	1	1	2	7	2	9
23	1	1	2	6	1	7
Total	9	21	30			

[a] $\Sigma\, m_{i1} = 21$; $\Sigma\, m_{i+} R_{i1}/R_{i+} = 10.749$; $\Sigma\, m_{i+} R_{i0} R_{i1}/R_{i+}^2 = 6.5957$; $U(0)/\sqrt{I(0)} = 3.992$

so that

$$U(0) = 10.251$$

and $\Sigma\, m_{i+} R_{i0} R_{i1}/(R_{i0} + R_{i1})^2 = 6.5957$ so that $I(0) = 6.5957$ and

$$U(0)/\sqrt{I(0)} = 3.992$$

indicating that the data are not consistent with the hypothesis that $\beta = 0$.

Multiple-Sample Problem

To compare p samples, we define $(p - 1)$ dummy regressors, $z_1, \ldots, z_{p-1}$, such that for the ith sample, z_i takes the value 1 and all other z's take the value 0, $i = 1, 2, \ldots, p - 1$, while for the pth sample all z's take the value 0. This amounts to treating the pth sample as the reference group. If the hazard function for the pth sample is $h_0(t)$, then that for the ith sample is

$$h_i(t) = h_0(t) \exp(\beta_i), \qquad i = 1, 2, \ldots, p - 1.$$

Notice that for the subjects in sample i, $\Sigma\, z_j \beta_j = \beta_i$.

To test the hypothesis that $\beta_1 = \cdots = \beta_{p-1} = 0$, we may use the score test analogous to the one illustrated above in the two-sample case. Let

$$t_1 < t_2 < \cdots < t_k$$

be the distinct failure times arranged in ascending order of magnitude, and let m_{ig} be the number of failures at t_i in sample g, and R_{ig} the number of individuals at risk just before t_i in sample g. Let m_{i+} stand for $m_{i1} + \cdots + m_{ip}$, and R_{i+} for $R_{i1} + \cdots + R_{ip}$. We construct a vector $U(0)$, whose gth element is

$$U_g(0) = \sum \left[m_{ig} - \frac{m_{i+} R_{ig}}{R_{i+}} \right]$$

and a matrix $I(0)$ whose (g, s) element is

$$I_{gs}(0) = \sum \left[m_{i+} \frac{R_{ig}}{R_{i+}} \left(\delta_{gs} - \frac{R_{is}}{R_{i+}} \right) \right]$$

where the summation is over i, and δ_{gs} takes the value 1 when $g = s$ and 0 otherwise. Under the hypothesis $\beta_1 = \cdots = \beta_{p-1} = 0$, the elements of $U(0)$ can be treated as jointly distributed as multivariate normal with mean $(0, \ldots, 0)$. To test the hypothesis that $\beta_1 = \cdots = \beta_{p-1} = 0$, we may use the statistic

$$X^2 = U'(0)[I(0)]^{-1} U(0)$$

treating it as a chi square with $(p - 1)$ degrees of freedom. (Large values of the statistic indicate that the data are not consistent with the hypothesis.) For further discussion see Cox (1972), Mantel (1966), Mantel and Haenszel (1959), Peto and Peto (1972), and Peto et al. (1977). One could also test hypotheses about linear combinations of the β's using the estimated variances and covariances of their estimates.

Example 1. Koo et al. (1984) examined several hypotheses concerning the effects of the number of children and the presence of a young child on the hazard rate of divorce after separation and of remarriage after divorce. Specific hypotheses selected for examination included

1. The hazard rate of divorce after separation is inversely related to the number of children and the presence of a young child at the time of separation.
2. The hazard rate of remarriage after divorce is inversely related to the number of children and the presence of a young child at the time of divorce.

The data used in the study were from the 1973 (U.S.) National Survey of Family Growth (NSFG). The regressors included in the analyses of the hazard rate on divorce after separation were

1. Number of children (0, 1, 2, 3 or more).
2. Age of youngest child (≤ 1, 2–5, 6 or above).
3. Birth cohort (1951–1959, 1946–1950, 1941–1945, 1936–1940, 1929–1935).
4. Region of residence in childhood (Northeast, North Central, South, West, other).
5. Parents' residence at age 14 (living with both parents, not so).
6. Education at marriage (0–8 years, 9–11 years, 12 years, more than 12 years).
7. Age at separation (20 or below, 21–25, 26–34, 35 or above).
8. Duration of marriage (under 4 years, 4–7 years, 8 or more years).
9. Religion (Catholic, other).

In the analysis of remarriage, to examine the interaction effects involving the first two variables (number of children and age of the youngest child), a new variable having ten categories was defined with no children as one category, and the three remaining categories of variable 1 crossed with the three categories of variable 2 as the other nine. The statistical significance of the main effects and interaction effects was tested under joint hypotheses of the type

$$H_0: \beta_1 = \cdots = \beta_{k-1} = 0$$

for effects represented by $k - 1$ nonredundant β's, e.g., the main effect of a k-category variable. [Note that tests of such hypotheses take into account the correlation structure of the estimators involved.] For whites and blacks the main effect of variable 1 was found significant. The main effect of variable 2, however, was found significant only for whites. The interaction involving the two variables was not significant in either group.

Example 2. Data from the 1980 U.S. Current Population Survey were used to estimate the following simple hazards model for remarriage for divorced women:

$$h(t \mid z) = h_0(t) \exp[\beta_1(\text{RACE}) + \beta_2(\text{AGEDIV1}) + \beta_3(\text{AGEDIV2})$$
$$+ \beta_4(\text{AGEDIV3})$$

where

RACE = 1 if black, and 0 if white

AGEDIV1 = 1 if divorced at age 20 or below, and 0 otherwise,

AGEDIV2 = 1 if divorced between ages 21 and 25, and 0 otherwise,

and

AGEDIV3 $= 1$ if divorced between ages 26 and 36, and 0 otherwise.

Note that for the AGEDIV variable "36 and above" was used as the reference category. The parameter estimates along with estimated standard errors are shown below.

Variable	$\hat{\beta}$: Estimate of regular coefficient	Estimate of standard error of $\hat{b}$	Column 1/Column 2	$\exp(\hat{\beta})$
RACE	−0.5310	0.0710	−7.4794	0.5880
AGEDIV1	0.3513	0.0673	5.2185	1.4209
AGEDIV2	0.2537	0.0587	4.3242	1.2888
AGEDIV3	−0.1121	0.0632	−1.7736	0.8939

The figures in the last column show clearly that blacks tend to remarry at a slower rate and that the duration pattern of remarriage is a decreasing one. To test the statistical significance of the model we use the likelihood ratio method. The negative of twice the likelihood evaluated at the estimated β's was 12120.0824. When all β's were set equal to zero, however, the corresponding figure was 12236.6426. The difference between the two figures, that is 116.56, can be treated as a chi square with 4 degrees of freedom. Thus the introduction of covariates significantly improves the fit.

Example 3. To provide an illustration of the use of time-dependent covariates, the following simple model was fitted to the same data used in Example 2:

$$h(t) = h_0(t) \exp[\beta_1(\text{RACE}) + \beta_2(\text{CONC}(t)) + \beta_3(\text{CONC}(t)(\text{RACE}))]$$

where $\text{CONC}(t) = 1$ if conception occurs before remarriage (after divorce) and 0 otherwise. Note that β_2 measures the effect of conception and β_3 the interaction effect involving conception and race. Both of these latter are time-dependent covariates in that their values change with time (duration since divorce). The results obtained by fitting the model are shown in the accompanying table.

Variable	$\hat{\beta}$	Standard error $(\hat{\beta})$	$\hat{\beta}$/Standard error $(\hat{\beta})$	$\exp(\hat{\beta})$
RACE	−0.5180	0.0791	−6.5502	0.5957
CONC	0.7427	0.0837	8.8787	2.1017
CONC × RACE	−0.5841	0.1726	−3.3837	0.5076

The improvement in the fit attributable to the covariates can be evaluated in terms of the likelihood ratio statistic comparing the maximum likelihood when all regressors are set equal to zero with the corresponding value when all regressors are retained. The test statistic turns out to be in the present case 188.22, which, treated as a chi square with three degrees of freedom, leads to the inference that the incorporation of the covariates significantly improves the fit of the model.

The hazard rate of remarriage, given conception, for whites is

$$\exp(0.7427) = 2.1017,$$

whereas the corresponding figure for blacks is $\exp(0.7427 - 0.5841) = 1.17$. Thus conception increases the risk of remarriage for whites but not for blacks.

3   Estimation of the Survival Function

Attention so far has been focused on the multipliers rather than the baseline hazard function $h_0(t)$. In most practical situations, interest centers on the impacts of covariates on the hazard rate, and inferences regarding these can be drawn without estimating the baseline hazard function. However, if one wishes to make specific reference to intergroup differences in survival functions, then the baseline hazard function, or equivalently the baseline survival function, must be estimated. Recall that

$$S(t; \mathbf{z}) = [S(t; 0)]^{\psi(\mathbf{z})} \qquad (12.25)$$

where $S(t; 0)$ is the survival function corresponding to the baseline hazard function; and $\psi(\mathbf{z}^*)$ is the multiplier for subjects with covariate score vector $\mathbf{z}^*$. Some of the methods developed for estimating $S(t; \mathbf{z})$ involve the following three steps. First, estimate the multipliers $\psi(\mathbf{z})$, which, it may be noted, can be estimated without estimating the baseline hazard function. Then estimate the baseline hazard function or the corresponding survival function. Finally use, if necessary, the relationship mentioned above to estimate $S(t; \mathbf{z})$. A procedure developed by Kalbfleisch and Prentice (1973) is described below.

To consider a simple setup, let

$$0 = t_0 < t_1 < t_2 < t_3 < t_4 = \infty$$

where t_i's are distinct failure times. Suppose individuals A, B, C, and D are reported to have failed, A and B at t_1, C at t_2, and D at t_3, and E is reported to have been censored at t_E, between t_2 and t_3. Using the notation ψ_i for the hazard multiplier for individual i, (i = A, B, ..., E), the contributions to the likelihood on the data from the observations are

$$[S_0(t_1)]^{\psi_A} - [S_0(t_1 + 0)]^{\psi_A} \qquad \text{from A}$$

$$[S_0(t_1)]^{\psi_B} - [S_0(t_1 + 0)]^{\psi_B} \qquad \text{from B}$$

$$[S_0(t_2)]^{\psi_C} - [S_0(t_2 + 0)]^{\psi_C} \qquad \text{from C}$$

$$[S_0(t_3)]^{\psi_D} - [S_0(t_3 + 0)]^{\psi_D} \qquad \text{from D}$$

and

$$[S_0(t_E)]^{\psi_E} \qquad \text{from E}$$

Consistent with the nonincreasing nature of the survival function, we now choose $S_0(t_1 + 0)$, $S_0(t_2 + 0)$, and $S_0(t_3 + 0)$ to be the least possible, $S_0(t_1)$ and $S_0(t_E)$ to be the highest possible so as to maximize the likelihood function. That is, we choose $S_0(t_1) = 1$, $S_0(t_1 + 0) = S_0(t_2)$, $S_0(t_2 + 0) = S_0(t_3)$, $S_0(t_3 + 0) = S_0(t_4) = 0$, and $S_0(t_E) = S_0(t_2)$. These choices make the contributions to the likelihood from A, B, C, D and E equal to

$$1 - [S_0(t_2)^{\psi_A}$$

$$1 - [S_0(t_2)]^{\psi_B}$$

$$[S_0(t_2)]^{\psi_C} - [S_0(t_3)]^{\psi_C}$$

$$[S_0(t_3)]^{\psi_D}$$

and

$$[S_0(t_2)]^{\psi_E}$$

respectively. The likelihood, i.e., the product of these contributions, simplifies to

$$(1 - p_1^{\psi_A})(1 - p_1^{\psi_B})[p_1^{\psi_C} - (p_1 p_2)^{\psi_C}][(p_1 p_2)^{\psi_D}]p_1^{\psi_E}$$

where $p_i = S_0(t_{i+1})/S_0(t_i)$, $i = 1$, 2, and 3. Equating to zero the partial derivative of the log likelihood with respect to p_1 gives

$$\frac{\psi_C + \psi_D + \psi_E}{p_1} = \frac{\psi_A p_1^{\psi_A}}{p_1(1 - p_1^{\psi_A})} + \frac{\psi_B p_1^{\psi_B}}{p_1(1 - p_1^{\psi_B})} \qquad (12.26)$$

which simplifies to

$$\psi_A + \psi_B + \psi_C + \psi_D + \psi_E = \frac{\psi_A}{1 - p_1^{\psi_A}} + \frac{\psi_B}{1 - p_1^{\psi_B}} \qquad (12.27)$$

the left-hand side of which is the sum of the multipliers for all those at risk as of time t_1, while the terms on the right-hand side pertain to those who fail at t_1.

Similarly, for p_2 we have the estimating equation

$$\psi_C + \psi_D = \frac{\psi_C}{1 - p_2^{\psi_C}} \tag{12.28}$$

This last equation illustrates the point that if there are no tied failure times at time t, then a closed form expression is available for the estimate of the corresponding value of p. If there are ties, however, no closed form expression is available [see (12.27)], and consequently iterative methods will have to be used for estimating the corresponding value of p. An initial approximation to p_1 can be obtained from (12.27) by replacing $p_1^{\psi_A}$ by $1 + (\ln p_1)\psi_A$ and $p_1^{\psi_B}$ by $1 + (\ln p_1)\psi_B$ and then solving for $\ln p_1$.

Example 1. Let us consider the data in Table 12.2 Using z to represent the sample ($z = 0$ for the control group, and $z = 1$ for the treatment group) and $\exp(z\beta)$ for the multiplier for the hazard function, we estimate β by maximizing the log likelihood on the data given by $L(\beta) = 21\beta - \Sigma\, m_{i1} \ln[R_{i0} + R_{i1} \exp(\beta)]$, where 21 is the total number of failures observed in the experimental group (corresponding to $z = 1$), m_{i1} is the number of failures observed at the ith failure time (ordered from the shortest to the longest), and R_{i0} and R_{i1} are the number of individuals at risk just before the ith failure time in samples 1 and 2, respectively. The required estimate is obtained by solving for β the equation

$$21 = \sum m_{i1} e^{\beta}/(m_{i0} + m_{i1} e^{\beta})$$

[One has to use iterative methods to solve the equation. The problem, however, is simple enough to permit trying out a few candidates for β (e.g., 0, 0.5, 1.5, 2.0) and then using interpolation to get closer to the solution.] The solution obtained by this latter method was 1.61. We now proceed to estimate the p values. The value of p_1 is obtained by solving the equation

$$2 \exp(\beta)/(1 - p_1^{\exp(\beta)}) = 21 + 21 \exp(\beta)$$

and the value of the survival function just after the first failure time is estimated as $p_0 p_1$, where $p_0 = 1$. Similarly, p_2 is estimated from the equation

$$2 \exp(\beta)/(1 - p_2^{\exp(\beta)}) = 21 + 19 \exp(\beta)$$

and the value of the survival function just after the second failure time is estimated as the product of p_0, p_1, and p_2, and so on. The survival function thus estimated is shown in Table 12.3.

Breslow (1974) suggests a simpler procedure based on the explicit formulas he was able to derive for the baseline hazard functions. Obviously if we have estimates of the baseline hazard function, we can immediately compute the values of the cumulative hazard function and from them obtain the

Table 12.3

Baseline Survival Function Estimated by Methods Proposed by Kalbfleisch and
Prentice (1973) and Breslow (1974)

t_i	$S_0(t_i + 0)$ Kalbfleisch and Prentice	Breslow	t_i	$S_0(t_i + 0)$ Kalbfleisch and Prentice	Breslow
1	0.98	0.98	11	0.80	0.80
3	0.97	0.96	12	0.77	0.77
4	0.96	0.95	13	0.73	0.73
5	0.94	0.93	15	0.71	0.71
6	0.92	0.91	16	0.68	0.68
7	0.88	0.88	17	0.66	0.66
10	0.82	0.82	22	0.63	0.63
			23	0.54	0.56

corresponding survival function using the formula $S(t) = \exp[-H(t)]$. The
baseline survival function estimated by Breslow's method is shown in
Table 12.3. Notice the closeness of the estimates.

Example 2. The survival functions estimated by the Breslow method for
the 1980 Current Population Survey data introduced earlier are plotted in
Fig. 12.1 for whites and blacks whose age at divorce was 20 years or younger.
Notice that the curve for blacks is above that for whites, indicating
the tendency for blacks to remain outside marriage longer than their white
counterparts, other things being equal.

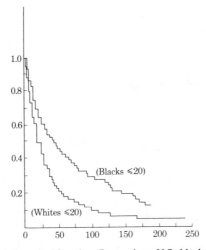

Fig. 12.1. Estimated survival function: Remarriage, U.S., blacks and whites, 1980.

4 Data Analysis Using the Proportional Hazards Model

The objective of survival data analysis may sometimes be strictly to test the statistical significance of a relationship hypothesized in advance of data collection. An example is the test of significance of a treatment effect in a well-designed clinical trial. But very often the analysis might be exploratory in nature. There may be, for example, a number of covariates (regressors) from which the most pertinent ones are to be identified. Sometimes the question may be whether the relationship between a regressor and the survival time conforms to the pattern implicated in a specified model. An example is the question whether a particular covariate acts upon the hazard rate in a multiplicative manner. A few data-analytic strategies that are helpful in dealing with such questions are mentioned below. For more thorough discussions of various aspects of regression analysis using proportional hazards model, the reader is referred to Kalbfleisch and Prentice (1980), Lawless (1982), Cox and Oakes (1984), Breslow (1975), and Kay (1977).

Coding of Variables for Inclusion in the Model

In many analyses, regressors are entered in the form of categorical variables (dichotomies, polytomies) either because of the categorical nature of the variables (e.g., religion, region of residence) or because specification as categorical variables permits examination of nonlinearity of effects. Categorical variables can be introduced in the model in various ways (Neter and Wasserman, 1974; Cohen and Cohen, 1983).

The Log Minus-Log Plot

The assumption of proportionality of hazards can be checked by inspecting the plots of $\ln(-\ln[S(t)])$. Suppose z_1 is a dichotomy taking values 1 and 2, say, and we wish to check whether z_1 affects the hazard function in accordance with the proportional hazards model:

$$h(t\,|\,\mathbf{z}) = h_0(t)\exp(\textstyle\sum z_j\beta_j)$$

Suppose $z_2, \ldots, z_k$ are the other covariates under consideration. We fit the model

$$h_1(t\,|\,\mathbf{z}^*) = h_{01}\exp(z_2 b_2 + \cdots + z_k b_k)$$

$$h_2(t\,|\,\mathbf{z}^*) = h_{02}\exp(z_2 b_2^* + \cdots + z_k b_k^*)$$

for individuals with $z_1 = 1$ and $z_1 = 2$, respectively. Suppose $s_{01}(t)$ and $s_{02}(t)$

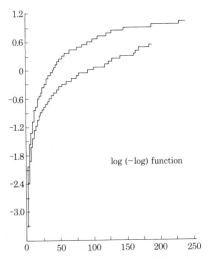

Fig. 12.2. Log minus log plots to check the tenability of the proportional hazards assumption: Remarriage, U.S. blacks and whites, 1980.

are the estimated baseline survival functions for those with $z_1 = 1$ and $z_1 = 2$, respectively. If z_1 acts upon the hazard function in accordance with the proportionality assumption, then $\ln(-\ln[s_{01}(t)])$ and $\ln(-\ln[s_{02}(t)])$ must have parallel plots. Departure from time-invariant separation of the two plots is an indication of a possible violation of the proportionality assumption. The log minus-log plots for whites and blacks in the 1980 *Current Population Survey* sample mentioned above (age at divorce being "fixed" at 20 or below) shown in Fig. 12.2 seem to indicate that the proportionality assumption is not too far off the mark.

Interaction Effects

If in the above setup, tests indicate that the regression coefficients in the two groups are not the same, that is, if there is indication that β_j and β_j^* are not the same ($j = 2, 3, \ldots, k$), it means that an interaction term involving z_1 and z_j should be included in the model.

Stratification

If a covariate does not affect the hazard function multiplicatively, then stratification may be used in the analysis. The model is specified as

$$h_j(t \mid \mathbf{z}) = h_{0j}(t)\exp(\textstyle\sum z\beta)$$

where each j represents one stratum. Under this model, the baseline hazard function is allowed to vary over strata, while the regression coefficients (β's) remain invariant from one stratum to another. Each stratum is treated separately when calculating the contributions to the likelihood. The product of the contributions from the various strata is maximized for estimation purposes. The baseline survival functions are estimated separately for each stratum.

Residual Analysis

Cox and Snell (1968) provide a general background for residual analysis. Particular application strategies with reference to ordinary regression models, are described in Draper and Smith (1981). In residual analysis, the issue examined is whether the estimated residuals indicate that the assumptions regarding residuals are violated. On the basis of our examination of the estimated residuals we declare that the underlying assumptions seem to have been violated or that the data indicate no violations. [The latter is not tantamount to saying that the assumptions made are indeed correct.] In general, the residuals are functions of the response variable, the covariates, and the parameters in the model specified. Thus, in the ordinary regression context, the residuals are functions of the dependent variable (response), the regressors, and the regression coefficients. They are ordinarily assumed to be independently and identically distributed according to a known distribution (e.g., normal in the ordinary regression context.) The estimated residuals (obtained by subtracting estimated response score from the corresponding observed score) are, of course, not independent of each other and are in general not identically distributed. But we usually treat them as though they are. Turning to the survival analysis context, consider the model

$$h(t \mid \mathbf{z}) = h_0(t) \exp(\textstyle\sum z_j \beta_j)$$

and recall that

$$S(t_i \mid \mathbf{z}) = \exp[-H(t_i \mid \mathbf{z})]$$

where $H(t_i \mid \mathbf{z})$ is the cumulative hazard function. It makes sense to regard

$$e_i = H(t_i \mid \mathbf{z})$$

as residuals for our analysis, because (1) $H(t_i \mid \mathbf{z})$ are functions of the response variable, the covariates, and the parameters, and (2) because $S(t_i \mid \mathbf{z}) = \exp[-H(t_i \mid \mathbf{z})]$ are independent random variables, uniformly distributed on $(0, 1)$, it follows that $H(t_i \mid \mathbf{z}) = -\ln[S(t_i \mid \mathbf{z})]$ are independent random vari-

Table 12.4

The Impact of Infant Death on the Intensity of Transition to the Next Higher Parity: Malaysia

| | | | | | | | | Year of birth | | | |
| | Education | | Ethnicity | | Boys among children living | Age of mother at birth of child | | 60s | More recent | nth child died | $(n-i)$th died, $i = 1, 2, \ldots$ |
Spacing	Husband	Wife	Chinese	Indian							
1–2	0.01	−0.02	0.46	0.41	—	−0.04		0.04	−0.00	−0.21	—
2–3	−0.01	−0.01	0.39	0.48	−0.07	−0.06		−0.02	−0.23	0.15	−0.02
3–4	0.00	−0.04	0.46	0.66	−0.38	−0.07		−0.05	−0.33	0.30	0.05
4–5	−0.01	−0.06	0.26	0.21	−0.13	−0.05		−0.10	−0.40	0.46	0.03
5–6	0.01	−0.07	0.18	0.20	−0.33	−0.06		−0.53	−0.65	−0.23	0.26
6–7	0.02	−0.03	−0.23	−0.02	0.45	−0.08		−0.22	−0.51	−0.01	0.12

ables identically distributed as standard exponentials. If the parameters are estimated by maximum likelihood methods, we may treat the corresponding estimated values of the cumulative hazard function for the uncensored observations as a random sample from a standard exponential, to a first approximation. For censored observations, a reasonable adjustment suggested in the literature (see, e.g., Lawless, 1982, p. 281) is to add 1 to each estimated value of the cumulative hazard function. If the model fits the data, plots of $-\ln[s(e)]$ against e, where $s(e)$ is the product-limit estimate of the survival function for the e's, should give approximately a straight line. The *BMDP* output provides such plots.

Problems and Complements

1. Discuss the types of covariates one usually encounters when analyzing infant mortality using hazard models. There are a number of computer programs now available to fit hazard models that incorporate different types of covariates. One such program is called RATE developed by Tuma *et al.* (1979). If covariates are all categorical, and the time dimension is also treated as categorical, one could use any of the programs available for log linear analysis. Discuss the use of the latter for fitting hazard models involving time-dependent covariates.

2. In the *World Fertility Survey*, information on contraceptive practice and breastfeeding was collected only for the open interval and the last closed interval. What bias results if one were to include in the analysis all open and closed intervals for which data on breast feeding and contraception were collected (Trussell *et al.*, 1985)?

3. Lehrer (1984) examined, on the basis of the *Malaysian Family Life Survey* data, whether the impact of infant death on child spacing varies by parity. The results of one preliminary analysis are shown in Table 12.4. Given that education was measured in number of years of schooling completed, that ethnicity was a trichotomy, with Malays as the reference category, that sex composition of surviving children was measured by the fraction of boys among surviving children, that calendar year of birth was treated as a trichotomy, with "before 1960" as the reference category, interpret the results presented in the table. First state what the figures say about the impact of the various covariates on the intensity of transition to the next higher parity; then indicate what the statements mean with respect to the impact of the covariates on the length of the interval between births.

Bibliographic Notes

The proportional hazards model was proposed by Cox (1972), with a log-linear form for the multiplier in the specification of the hazard function. Thomas (1981) has suggested a linear form instead, while Efron (1977) has proposed a logistic form. Kalbfleisch and Prentice (1973) interpreted Cox's likelihood in terms the marginal likelihood of ranks. For grouped data, the marginal likelihood was derived by Kalbfleisch and Prentice (1973). Prentice and Gloeckler (1978) derived the unconditional analysis of grouped data, with a separate parameter for each fixed, grouping interval.

Estimators for the baseline hazard function were suggested by Cox (1972), Breslow (1974, 1975), and others. Many computer programs use Breslow's (1974) version of Cox's regression.

Mantel (1966) derived the log rank test for comparing survival functions, using Mantel and Haenszel's (1959) test as an analogy. Peto *et al.* (1977) provide a nontechnical exposition of the log rank test.

Crowley and Hu (1977) present an interesting example of Cox's model involving time-dependent covariates. For a comparison of Cox and Weibull models see Byar (1983). Fitting Cox's model using GLIM has been discussed by Whitehead (1980).

The notion of generalized residuals is Cox and Snell's (1968). Crowley and Hu (1977) show how to use the notion in survival data analysis. Kay (1977) also provides an illustrative application of the notion. Lagakos (1981), however, cautions their interpretation.

Chapter 13　Parametric Regressions

In Chapter 12 it was mentioned that the dependency of failure time on explanatory variables can be analyzed by considering the hazard rate or the failure time itself as the dependent variable. The focus of Chapter 12 was on the former approach, with the hazard rate expressed as a product of an arbitrary baseline hazard pattern and a multiplier representing the effects of explanatory variables. It was mentioned then that one could replace the arbitrary part with the hazard function of known distributions, thus producing a fully parameterized representation of the hazard rate. We now consider other ways to represent the dependency in question using completely parameterized models.

First let us look at some examples. Nelson (1970) used the conjecture, based on engineering considerations, that the time to failure of an electric insulating fluid under constant voltage stress has a Weibull distribution, with stress differences affecting the scale parameter rather than the shape parameter. To give another example, Feigl and Zelen (1965) hypothesized that the probability of surviving a specified length of time for patients with acute myelogenous leukemia follows an exponential law, with the white blood cell count at the time of diagnosis serving as a covariate (there being an inverse relationship between the probability of surviving a given length of time and the white blood cell count). Similarly, Byar *et al.* (1974) fitted an exponential model for time to death of patients with advanced prostate cancer. The mean of the distribution was hypothesized to be a function of (1) presence of pain due to the particular cancer, (2) acid phosphates in King Armstrong (K. A.) units, (3) presence of ureteral dilation, (4) metastasis (osseous or soft part), (5) physical mobility, (6) weight, (7) hemoglobin level, and (8) age.

From these examples it should be clear that based on previous research, theory, or other considerations, one may specify the dependence of failure time on explanatory factors by expressing one or more parameters of a known probability distribution as a parametric function of the explanatory factors. Let us examine the use of a few specific distributions in this context.

1 Exponential Distribution

Recall that the exponential distribution has the probability density function

$$f(t) = \theta e^{-\theta t} \tag{13.1}$$

the mean of the distribution being $1/\theta$. If we transform the random variable T having pdf (13.1) to its natural logarithm, say, Y, we see that Y has the pdf given by

$$f(y) = \exp(\epsilon - e^{\epsilon}) \tag{13.2}$$

where

$$\epsilon = y - \lambda \quad \text{or} \quad y = \lambda + \epsilon \tag{13.3}$$

and

$$\theta = \exp(-\lambda) \tag{13.4}$$

Thus, if failure time follows an exponential distribution, with pdf (13.1), then the natural logarithm of failure time has an extreme-value distribution, with pdf (13.2). Suppose we now reparameterize λ in (13.4) as $\Sigma\, z_j\beta_j$, where the z_j's are explanatory factors. Then we have the following model for the natural logarithm of failure time:

$$y = \sum z_j\beta_j + \epsilon \tag{13.5}$$

where ϵ has the standard extreme value distribution. Thus, if the sample is from an exponential distribution, we might express the dependence of failure time on explanatory factors in terms of either a reparameterization of the exponential model itself, or a linear-regression type dependence of the logarithm of failure times on the explanatory factors. This latter approach is different from the one discussed in Chapter 11 because here we do not require that all the explanatory factors be defined as categorical variables. It should be noted that various ways of reparameterizing θ of (13.1) other than the one mentioned above have been suggested in the literature. Thus, for example, Zippin and Armitage (1966) suggest the reparameterization $\theta = \Sigma\, z_j\beta_j$, and Greenberg *et al.* (1974) suggest instead $\theta = (\Sigma\, z_j\beta_j)^{-1}$. The advantage of (13.5) is that the requirement $\theta > 0$ is automatically satisfied for all z and β, whereas with the other two reparameterizations this requirement calls for placing restrictions on the β's. We consider only the form (13.5).

2   Maximum Likelihood Methods

The likelihood on the data $t_i, z_{i1}, \ldots, z_{ik}, i = 1, 2, \ldots, n$, is

$$\text{lik}(\beta_1, \ldots, \beta_k) = \prod \exp[\epsilon_i - \exp(\epsilon_i)] \prod \exp[-\exp(\epsilon_i)]$$

where the first product is over the uncensored cases and the second is over the censored times. The log likelihood on the data for the β's is

$$L = \sum_u \epsilon_i - \sum \exp(\epsilon_i) \tag{13.6}$$

where the first summation is over the uncensored cases and the second is over all cases, and

$$\epsilon_i = y_i - \sum_j z_{ij}\beta_j \tag{13.7}$$

The maximum likelihood estimates of the β's are given by the solutions of the simultaneous system of equations

$$\partial L/\partial \beta_j = 0, \qquad j = 1, \ldots, p \tag{13.8}$$

These equations are solved by standard iterative methods taking into account the second derivatives of the log likelihood. If censoring is not heavy, initial estimates can be obtained by the least-squares method, treating the censored cases as though they were uncensored. Approximate tests for the β's can be carried out on the basis of likelihood ratios or normal approximations for the estimators of the β's, the latter being quite a reasonable approach if the sample size is not "too small" (e.g., not less than 25), according to a study by Singhal (1978) cited in Lawless (1982). Convergence in the iterative method used in this connection is often quicker if the values of the explanatory variables are entered as deviations from the respective observed means.

Example 1. In 1960, the Veteran Administration's Cooperative Urological Research Group (VACURG) undertook a study designed to compare several standard methods of treatment for cancer of the prostate. In 1967, the group analyzed the survival experiences of patients through August 15, 1966. Greenberg *et al.* (1974) reanalyzed the data to estimate the effects of eight factors. All except one of the eight factors were defined as categorical variables: Age (75–79, 80+, other); hemoglobin (< 12.0 g/100 cm^3, other); weight (< 130 lbs., other), physical mobility (partially confined to bed, totally confined to bed, other), metastases (present, absent); ureteral dilation (present, absent), pain due to prostate cancer (present, absent), acid phosphate (actual measurement if ≤ 5 K.A. units, 5 otherwise).

The analysis was confined to patients who were originally assigned to placebo; the sample size was 302. The procedure used was one of backward elimination: starting with all eight regressors, deletion resulted in identifying metastases and acid phosphate as the only two significant regressors for retention in the model. Based on the fitted model, the authors computed expected number of deaths in each of 20 time intervals, the first 19 of which were of length 3 months and the last one over 57 months. The goodness of fit of the model was assessed by comparing expected counts thus computed with the corresponding observed counts. The statistic $X^2 = \Sigma(\text{observed-expected})^2/\text{expected}$ was treated as a chi square with 16 degrees of freedom. From the computed value, 23.27, of X^2, the inference was drawn that the fit was reasonably good.

Example 2. Krall *et al.* (1975) report survival times and values of several regressors (logarithm of blood urea nitrogen, hemoglobin measurement, age, sex, and serum calcium) for multiple myelomous patients. They used a step-up procedure for selecting significant regressors for retention in the model. Likelihood ratio method was used in deciding whether to retain a regressor. The authors examine the validity of the assumption that the likelihood ratio statistic used in the analysis has a chi square distribution on the basis of simulated data.

Piecewise Exponential. There is no new principle involved in carrying over to the present context the ideas described in Chapter 11 with reference to piecewise exponential models. This approach, when the regressors are not categorical, however, remains to be fully taken advantage of in empirical analyses.

3 Weibull Regression Models

The Weibull distribution has the pdf

$$f(t) = \lambda\alpha(\lambda t)^{\alpha-1} \exp[-(\lambda t)^\alpha] \tag{13.9}$$

where $\alpha > 0$ and $\lambda > 0$, are parameters. The latter determines the shape of the distribution and hence is called the shape parameter; the former is called the scale parameter because it determines the scale on the horizontal axis. The corresponding survival function is

$$S(t) = \exp[-(\lambda t)^\alpha] \tag{13.10}$$

It has already been pointed out that the flexibility of this distribution permits its use in describing the patterns of a wide variety of survival data. The distribution is a good candidate for use in expressing dependence of failure time on explanatory factors. A very common procedure is to let the scale parameter be a function of explanatory factors, leaving the shape parameter untouched. Transforming the random variable T (representing failure time) to its logarithm, say, Y, we have for the pdf of the latter

$$f(y) = \alpha(\lambda e^y)^\alpha \exp[-\lambda e^y] \tag{13.11}$$

Now let us replace λ with $\exp(-\Sigma\, z_j\beta_j)$. Then (13.11) becomes

$$\alpha \exp\{\alpha(y - \sum z_j\beta_j) - \exp[\alpha(y - \sum z_j\beta_j)]\} \tag{13.12}$$

On right-censored data, the log likelihood for the parameters has the form

$$L = m \ln \alpha + \sum_u \alpha[y_i - (z_{i1}\beta_1 + \cdots + z_{ik}\beta_k)]$$

$$- \sum \exp\{\alpha[y_i - (z_{i1}\beta_1 + \cdots + z_{ik}\beta_k)]\} \tag{13.13}$$

where m stands for the number of failures, Σ_u denotes the summation over failure times, and Σ denotes summation over all observations, i.e., failure times and censorings. There are $(k + 1)$ parameters to be estimated. To obtain maximum likelihood estimates, one solves simultaneously the equations

$$\partial L/\partial \alpha = 0 \quad\text{and}\quad \partial L/\partial \beta_j = 0, \quad j = 1, \ldots, k \tag{13.14}$$

Iterations usually converge faster if the regressors are entered as deviations from their respective means. For testing purposes likelihood ratio methods or the approximate normality of the maximum likelihood estimators may be used. Model selection can be attempted via the backward elimination or the forward inclusion (step-up) procedures. One might also try the procedure involving the examination of all possible regressions. There is no guarantee that these methods will yield identical results. Lawless (1982) discusses Weibull regression models at length (also see Farewell and Prentice, 1977).

4 Log-Normal Regression Models

As mentioned in Chapter 10, if the natural logarithm of survival time is normally distributed, we say that the survival time has a log-normal distribution. Consider for example survival time of breast cancer patients. Suppose we have data on the patient's age, node status (clinically negative or palpable), tumor size in cm, and degree of skin fixation (none, incomplete, complete). Let us assume that the patients are otherwise similar to each other.

We consider the following regressors: $z_0 \equiv 1$; $z_1 =$ observed age $-$ mean age for the sample, $z_2 =$ observed tumor size $-$ the corresponding sample mean, $z_3 = 1$ if node status is palpable and 0 otherwise, and z_4 and z_5 are defined as follows:

z_4	z_5	
1	0	If skin fixation is incomplete
0	1	If skin fixation is complete
-1	-1	If no skin fixation

We consider the normal regression model

$$y_i = \sum z_{ij}\beta_j + e_i \qquad (13.15)$$

where we assume that e_i's are normally distributed with mean 0 and standard deviation σ, and are independent of each other. In (13.15) the pdf and survival function of Y_i, given z's, are

$$f(y_i \,|\, z\text{'s}) = \frac{1}{\sigma}\,\phi\!\left[\frac{y_i - \sum z_{ij}\beta_j}{\sigma}\right] \qquad (13.16)$$

$$S(y_i \,|\, z\text{'s}) = Q\!\left[\frac{y_i - \sum z_{ij}\beta_j}{\sigma}\right] \qquad (13.17)$$

where $\phi(x)$ and $Q(x)$ are the standard normal pdf and survival function, the latter being the area under the standard normal curve to the right of x.

Suppose that, in a random sample of n individuals, m log failure times and $n - m$ log censoring times have been observed (y_i may stand for log failure time or log censoring time). For convenience, let D and C, respectively, stand for the label set of individuals for whom failure times and censoring times have been observed. Then the likelihood on the data for the parameters involved is

$$\text{lik} = \prod_{i \in D} \frac{1}{\sigma}\,\phi\!\left[\frac{y_i - \sum z_{ij}\beta_j}{\sigma}\right] \prod_{i \in C} Q\!\left[\frac{y_i - \sum z_{ij}\beta_j}{\sigma}\right] \qquad (13.18)$$

and the log likelihood is

$$L = -m \ln \sigma - \frac{1}{2\sigma^2} \sum_{i \in D}(y_i - \sum z_{ij}\beta_j)^2 + \sum_{i \in D} \ln Q\!\left[\frac{y_i - \sum z_{ij}\beta_j}{\sigma}\right] \qquad (13.19)$$

If we write $V(x)$ for $\phi(x)/Q(x)$, and u_i for $y_i - \sum z_{ij}\beta_j$, the equations we need to solve simultaneously in order to obtain the maximum likelihood estimates of

σ and the β's can be expressed as

$$\frac{\partial L}{\partial \beta_j} = \frac{1}{\sigma} \sum_{i \in D} z_{ij} u_i + \frac{1}{\sigma} \sum_{i \in C} z_{ij} V(u_i) = 0, \qquad j = 1, \ldots, k \qquad (13.20)$$

$$\frac{\partial L}{\partial \sigma} = -\frac{m}{\sigma} + \frac{1}{\sigma} \sum_{i \in D} u_i^2 + \frac{1}{\sigma} \sum_{i \in C} u_i V(u_i) = 0 \qquad (13.21)$$

These equations can be simultaneously solved using iterative methods. If censoring is heavy, it is known that some techniques such as the Newton–Raphson iteration may fail to converge, and when they do the result may sometimes be a negative estimates for σ (Lawless, 1982). A procedure developed by Sampford and Taylor (1959) has been adapted by Wolynetz (1979) for use in such situations. Once we know how to obtain maximum likelihood estimates and are able to calculate the likelihood for given parameter values, the rest of the inference procedure is rather straightforward. Tests of significance can be carried out using likelihood ratio methods or the asymptotic normality of maximum likelihood estimators.

5   Log-Logistic Regression Models

The log-logistic distribution has a nonmonotonic hazard function which makes it suitable for modeling certain types of cancer survival data, such as those reported by Langlands *et al.* (1979) in a study of the curability of breast cancer. [In that study peak mortality occurred after about three years.] The log-logistic distribution is very similar to the log-normal distribution, but mathematically the former is more tractable when there is censoring. The log-logistic distribution function, that is, $1 - S(t)$, for a single sample is

$$F(t; \varphi, \theta) = [1 + t^{-\varphi} e^{-\theta}]^{-1} \qquad (13.22)$$

where θ is a measure of location and $\varphi > 0$ is a measure of dispersion. For a regression model one reparameterizes θ by a parametric function of the covariates, e.g., $\Sigma \, z_{ij} \beta_j$ for individuals in the ith sample. Using the notation $F(t)$ for the distribution function defined above, we notice that

$$\ln\{F(t)/[1 - F(t)]\} = \varphi \ln t + \theta$$

The expression on the left-hand side is the log odds on survival time. Clearly, if φ remains invariant, say from sample to sample, then differences between samples in log odds on survival time are reflected in the corresponding differences in θ. Thus in a two sample problem, with one covariate taking value 0 for one sample and 1 for the other, the difference in the log odds on survival to t is simply β, which is time invariant. Thus one odds function is a constant multiple, $\exp(\beta)$ of the other. Bennett (1983) describes how to use GLIM (Baker and Nelder, 1978) to fit log-logistic regression models. In a multisample problem, the hazard function for the sth sample is given by

$$h_s(t) = (\varphi/t)[1 + t^{-\varphi} \exp(-\psi_s)]^{-1}$$

where ψ_s is a linear parametric function of covariates $z_1, \ldots, z_k$. It has its maximum at $t = \{(\varphi - 1) \exp(-\psi_s)\}^{1/\varphi}$. The hazards ratio of two samples converges to unity as t increases. Thus the hazards functions for different samples are not proportional. The two properties just mentioned are desirable, for example, when differences between two treatment effects are suspected to diminish over time. For the transformed variable $Y = \ln T$, the logistic distribution has pdf

$$f(y) = \frac{\varphi \exp(-\varphi y - \theta)}{[1 + \exp(-\varphi y - \theta)]^2} \tag{13.23}$$

The corresponding distribution function $= (1 - \text{survival function})$ is

$$F(y) = [1 + \exp(-\varphi y - \theta)]^{-1} \tag{13.24}$$

When working with logarithms of the observed time, it should be recognized that small observations tend to have, in general, somewhat disproportionate effect on the goodness of fit of models. Bennett (1983) suggests that this difficulty can be overcome by left-censoring all times that fall below a certain chosen value. If we have left- and right-censored observations in the sample, the contribution of the former is the corresponding value of $F(y)$, and that of the latter the corresponding value of $1 - F(y)$. [For an uncensored observation, of course, the contribution is the corresponding value of $f(y)$.] For notational convenience if we write p_{iy} for the value of $F(y)$ in the ith sample, then noting that the corresponding $f(y) = \varphi p_{iy}(1 - p_{iy})$, the individual contributions to the likelihood from uncensored, left-censored, and right-censored observations in the ith sample are, respectively, $\varphi p_{iy}(1 - p_{iy})$, p_{iy}, and $[1 - p_{iy}]$. The likelihood on the data is then the product of these contributions over all observations in all the samples. As an illustration we turn to the data from the Veteran's Administration lung cancer trial (Prentice, 1973), in which males with advanced inoperable lung cancer received chemotherapy. The data for the subsample of patients ($n = 97$) who received no prior therapy was analyzed by Bennett (1983) using the method described above. Two covariates were considered: tumor type (large, adeno, small, squamous); and performance status, a measure of general fitness on a scale of 0 to 100. (Farewell and Prentice (1977) fit Weibull, log-normal, and log-gamma regression models to the same subset of patients and the same pair of covariates.) Parameter estimates reported by Benette include the following:

Effect of performance status: -0.055, with a standard error of 0.010;
Effect of tumor type (with "large" tumor as the reference category):

adeno	1.295 (standard error, 0.552)
small	1.372 (standard error, 0.521)
squamous	-0.138 (standard error, 0.576)

The estimated odds ratios are: $\exp(1.295)$ or 3.651 for adeno versus large,
$\exp(1.372)$ or 3.943 for small versus large.

6   Other Fully Parameterized Regression Models

Reference may be made to Lawless (1982) for a detailed discussion of fully
parameterized models involving the distributions mentioned above as well as
others such as gamma and log-gamma distributions. Aitkin and Clayton
(1980) show how to fit exponential, Weibull, and extreme-value distributions
to complex censored survival data using GLIM (Baker and Nelder, 1978).
They express the likelihood in each case as a Poisson likelihood, with a log-
linear model for the Poisson mean corresponding to the log-linear model for
the hazard function. This permits the application of the program for
categorical data analysis in fitting regression models to data from the
distribution mentioned. Their reasoning can be summarized as follows. Let t_i,
$i = 1, \ldots, n + m$, be the survival times of $n + m$ individuals, the last m of which
are right censored. Let z_{ij}, for $i = 1, \ldots, n + m$, and $j = 0, 1, \ldots, k$ be the
corresponding values of explanatory variables, with $x_{i0} = 1$. Let the hazard
function be of the form $h(t_i) = \theta(t_i) \exp(\Sigma_j \beta_j z_{ij})$. Then if $\Theta(t)$ stands for the
area to the left of t of the $\theta(u)$ curve, and δ_i is an indicator variable taking the
value 1 for uncensored and 0 for censored observations, then the likelihood on
the data is

$$\text{lik} = \prod_{i}^{n+m} [\mu_i^{\delta_i} \exp(-\mu_i)][\theta(t_i)/\Theta(t_i)]^{\delta_i} \qquad (13.25)$$

where $\mu_i = \Theta(t_i) \exp(\Sigma \beta_j z_{ij})$. The product of the factors on the left is the kernel
of the likelihood function of $(n + m)$ independent "Poisson variates" δ_i with
means μ_i. The remaining portion of the likelihood does not contain any β's but
may contain other parameters. Thus in the case of an exponential distribution,
$\theta(t)$ being constant at, say, θ, $\Theta(t) = \theta \times t$. Consequently,

$$\ln \mu_i = \ln \theta + \ln t_i + \Sigma \beta_j z_{ij}$$

and $\theta(t_i)/\Theta(t_i) = t_i$. In the case of a Weibull distribution with $t \geq 0$, and pdf given by (13.9), $h(t)/H(t) = \alpha/t$, which depends on the unknown shape parameter α, and must be estimated along with β's. The likelihood equations (13.14) are (with all summations being from 1 to $n + m$)

$$\frac{\partial L}{\partial \beta_j} = \sum (\delta_i - \mu_i) z_{ij} = 0$$

$$\partial L/\partial \alpha = \frac{n}{\alpha} + \sum (\delta_i - \mu_i) \ln t_i = 0$$

showing that the maximum likelihood estimate of α satisfies the equation

$$\hat{\alpha} = \left[\sum (\hat{\mu}_i - \delta_i) \times (\ln t_i)/n \right]^{-1} \tag{13.26}$$

To estimate simultaneously α and the β's the iterative procedure starts with setting $\alpha_0 = 1$, uses the algorithm for fitting log-linear models to fit β's, then estimates α from (13.26) and repeats the process until convergence. Aitkin and Clayton (1980) note that after obtaining two estimates for α, convergence would be faster, if their average is used rather than the most recent one in the next cycle. Once maximum likelihood estimates of the parameters have been obtained, the maximized likelihood can be evaluated and used in likelihood ratio tests. Model selection proceeds in the same way as in categorical data analysis (see Aitkin, 1978).

7 A General Approach

An approach general enough to cover several of the methods mentioned above as particular cases has been recently suggested by Clayton (1983). The trick is to start with a specification of the baseline hazard function in a general but not arbitrary form. One such form is a positive transformation of a polynomial in survival time or its logarithm. Thus, suppose we specify

$$h(t) = \beta_0 + \beta_1 t + \beta_2 t^2 + \cdots \tag{13.27}$$

or

$$h(t) = \exp(\beta_0 + \beta_1 t + \beta_2 t^2 + \cdots) \tag{13.28}$$

Clearly, $h(t) = \beta_0$ is a special case of (13.27), and this in turn corresponds to an exponential distribution. Similarly, $h(t) = \exp(\beta_0 + \beta_1 t)$ giving Gompertz' distribution is a special case of (13.28). Suppose we insert the natural logarithm of survival time instead of t in (13.28). Then we have, writing y for $\ln t$,

$$h(t) = \exp[\beta_0 + \beta_1 y + \beta_2 y^2 + \cdots]$$

giving $h(t) = \exp[\beta_0 + \beta_1 y]$, the hazard function of a Weibull distribution as a particular case.

Consider n observations at times $t_1 \leq t_2 \leq \cdots \leq t_n$, each observation being a failure time or a right-censored time. Defining $\delta_i = 1$ if t_i is a failure time and 0 if it is not, the likelihood on the data may be expressed as

$$\text{lik} = \prod_{i=1}^{n} \left\{ [h(t_i)]^{\delta_i} \exp\left[-\int_0^{t_i} h(u)\, du \right] \right\} \tag{13.29}$$

which is a function of the data and the hazard function $h(t)$. Clayton (1983) suggests that the integral in (13.29) be evaluated numerically using the trapezoidal rule. Writing for convenience h_i for the value of the hazard function at $t = t_i$, the trapezoidal approximation using the values of $h(t)$ at the observed failure and censored times gives

$$-\ln[S(t_i)] = \int_0^{t_i} h(u)\, du = \frac{1}{2} \sum_{j=1}^{i} (h_j + h_{j-1})(t_j - t_{j-1}) \tag{13.30}$$

where $t_0 = 0$ is the origin for survival times. On substitution in (13.29) we obtain the approximation for the likelihood on the data

$$\text{lik} \simeq \prod_i [h_i^{\delta_i} \exp(-k_i h_i)] \tag{13.31}$$

where

$$\delta_0 = 0$$

and

$$2k_0 = R_1(t_1 - t_0)$$
$$2k_i = R_i(t_i - t_{i-1}) + R_{i+1}(t_{i+1} - t_i), \qquad i = 1, 2, \ldots, n-1$$
$$2k_n = R_n(t_n - t_{n-1})$$

R_i being the number of observations surviving to just before t_i. The log likelihood corresponding to (13.31) is

$$L = \sum_{i=0}^{n} [\delta_i \ln h_i - k_i h_i] \tag{13.32}$$

Notice that h_i is a function of the parameters involved in the specification of $h(t)$. The parameters are estimated by maximizing L. Computer programs available in SAS, BMDP, and other packages can be used for maximization of L. To apply the procedure one starts with a specification of the hazard function as a positive transformation of a polynomial in t or some positive function of t (e.g., $\ln t$). Values of $h(t)$ at the observations are then expressed in terms of the parameters of the model. Log likelihood is specified in terms of these, and the parameters are estimated by maximizing the log likelihood.

Thus far, attention has been confined to the baseline hazard function. To incorporate explanatory variables one proceeds in the usual manner, e.g., by specifying the hazard function as a product of the baseline hazard function and a multiplier that is a function of the explanatory variables.

8   Model Life Tables

The idea of expressing the survival function or hazard function of one population as a modification of a baseline survival or hazard function characteristic of a reference population has a parallel in the study of mortality tables of human populations. Brass *et al.* (1968) propose that if the probabilities of survival to age x were transformed to their logits, then, in the transformed scale, the survival function of any population can be expressed, to a close approximation, as a linear transformation of the survival function of a standard population. Thus if $l_s(x)$ denotes the survival function of a standard population, and $l(x)$ the survival function of any population, then, to a close approximation

$$\lambda(x) = \alpha + \beta \lambda_s(x) \tag{13.33}$$

where $\lambda(x)$ and $\lambda_s(x)$ are the logit transforms of $l_s(x)$ and $l(x)$, respectively. That is,

$$\lambda(x) = 0.5 \ln[l(x)/(1 - l(x))]$$

and $\lambda_s(x)$ is similarly defined. Brass proposes two standard populations, one called the general standard and the other the "African" standard, the latter reflecting relatively high childhood mortality. [Other scholars have identified more than two "standards," more commonly known as model life tables. See, for example, Coale and Demeny (1966, 1983), Ledermann (1969), United Nations (1955). In explanatory studies, we might use various population traits to account for the variation in the α's and β's that characterize mortality patterns under a given standard.

Bibliographic Notes

Lawless (1982) has devoted one whole chapter of his book to parametric regressions. Applications of the exponential regression model can be found in Aitkin and Clayton (1980), Fiegl and Zelen (1965), Krall *et al.* (1975), Prentice (1973), and Zelen (1959). Zelen (1959, 1960) and Lawless and Singhal (1980) have examined models of the type used in analysis of variance for factorial life testing experiments, with emphasis on the comparison of maximum-likelihood and least-squares estimates.

Reports on the application of Weibull regression models are available in Pike (1966), Peto and Lee (1973), and Nelson (1972, 1982). Roger and Peacock (1982) discuss the fitting of the scale as a GLIM parameter for Weibull, extreme-value, logistic, and log-logistic regression models with censored data.

Normal and log-normal regression models have been applied in the works of Feinleib (1960), Glasser (1965), Nelson and Hahn (1972, 1973), and Whittemore and Altschuler (1976).

The sensitivity of inferences to violations of assumptions remains to be more thoroughly investigated.

References

Aalen, O. (1976). Nonparametric inference in connection with multiple decrement models. *Scand. J. Stat.* **3**:15–27.

Adlakha, A. (1972). Model life tables: An empirical test of their applicability to less developed countries. *Demography* **9**:589–601.

Aitchison, J., and Brown, J. A. C. (1957). "The Lognormal Distribution." Cambridge Univ. Press, London and New York.

Aitkin, M. (1978). The analysis of unbalanced cross-classifications (with discussion). *J. R. Stat. Soc. Ser. A* **141**:195–223.

Aitkin, M., and Clayton, D. G. (1980). The fitting of exponential, Weibull, and extreme value distributions to complex censored survival data using GLIM. *Appl. Stat.* **29**:156–163.

Aitkin, M., and Wilson, G. T. (1980). Mixture models, outliers, and the EM algorithm. *Technometrics* **22**:325–331.

Alderson, M. (1981). "International Mortality Statistics." Facts on File, Inc., New York.

Altschuler, B. (1970). Theory for the measurement of competing risks in animal experiments. *Math. Biosci.* **6**:1–11.

Anscombe, F. J. (1961). Estimating a mixed-exponential law. *J. Am. Stat. Assoc.* **56**:493–502.

Armitage, P. (1959). The comparison of survival curves. *J. R. Stat. Soc. Ser. A* **122**:279–300.

Armitage, P. (1975). "Sequential Medical Trials," 2nd Ed. Wiley, New York.

Bain, L. J. (1972). Inferences based on censored sampling from the Weibull or extreme value distribution. *Technometrics* **14**:693–702.

Bain, L. J. (1974). Analysis for the linear failure rate distribution. *Technometrics* **16**:551–559.

Bain, L. J. (1978). "Statistical Analysis of Reliability and Life-Testing Models." Dekker, New York.

Bain, L. J., and Engelhardt, M. (1980). Probability of correct selection of Weibull versus gamma based on likelihood ratio. *Commun. Stat. A* **9**:375–381.

Baker, R. J., and Nelder, J. A. (1978). "General Linear Interactive Modeling (GLIM)." Release 3, Numerical Algorithm Group, Oxford, England.

Barlow, R. E., and Proschan, F. (1975). "Statistical Theory of Reliability and Life Testing." Holt, New York.

Barlow, R. E., and Singpurwalla, N. D. (1975). "Statistical Theory of Reliability and Life Testing." Holt, New York.

Barnett, V. (1975). Probability plotting methods and order statistics. *Appl. Stat.* **24**:95–108.

Bartholomew, D. J. (1957). A problem in life testing. *J. Am. Stat. Assoc.* **52**:350–355.

Basu, A. P., and Gosh, J. K. (1978). Identifiability of the multinormal and other distributions under competing risks model. *J. Multivariate Anal.* **8**:413–429.

Batten, R. W. (1978). "Mortality Table Construction." Prentice-Hall, Englewood Cliffs, New Jersey.

Benjamin, B., and Haycocks, H. W. (1970). "The Analysis of Mortality and Other Actuarial Statistics." Cambridge Univ. Press, London and New York.

Bennett, S. (1983). Analysis of survival data by the proportional odds model. *Stat. Med.* **2**.

Berkson, J., and Elvebeck, (1960). Competing exponential risks, with particular reference to smoking and lung cancer. *J. Am. Stat. Assoc.* **55**: 415–428.

Berretoni, J. N. (1964). Practical applications of the Weibull distribution. *Ind. Qual. Control* **21**:71–79.

Bhat, U. N. (1984). "Elements of Applied Stochastic Processes," 2nd Ed. Wiley, New York.

Billman, B., Antle, C., and Bain, L. J. (1972). Statistical inferences from censored Weibull samples. *Technometrics* 14:831–840.

Birnbaum, Z. W. (1979). On the mathematics of competing risks. *Vital Health Stat. Ser.* 2 (77): 1–77. U. S. National Center for Health Statistics.

Bishop, Y. M. M., Fienberg, S. E., and Holland, P. W. (1975). "Discrete Multivariate Analysis." MIT Press, Cambridge, Massachusetts.

Blalock, Jr., H. M. (1964). "Causal Inferences in Nonexperimental Research." Univ. of North Carolina Press, Chapel Hill, N. C.

Boag, J. W. (1949). Maximum likelihood estimates of the proportion of patients cured by cancer therapy. *J. R. Stat. Soc. Ser. B* 11:15–53.

Böhmer, P. E. (1912). Theorie der unabhängigen Wahrscheinlichkeiten Rapports. "Mémoires et Procès-verbaux de Septième Congrès International d'Actuaries," Amsterdam, Vol. 2, pp. 327–343.

Bongaarts, J., Burch, T. K., and Wachter, K. W., eds. (1985). "Family Demography: Methods and Their Application." Oxford Univ. Press, London and New York.

Brass, W. (1958). The distribution of births in human populations. *Popul. Stud.* 12:51–72.

Brass, W. (1974). Perspectives in population prediction: Illustrated by the statistics of England and Wales. *J. R. Stat. Soc. Ser. A.* 137, 532–583.

Brass, W., Coale, A. J., Demeny, P., Heisel, D. F., Lorimer, F., Rumanink, A., and van de Walle, E. (1968). "The Demography of Tropical Africa." Princeton Univ. Press, Princeton, New Jersey.

Braun, H. I. (1980). Regression-like analysis of birth interval sequence. *Demography* 17:207–223.

Breslow, N. E. (1970). A generalized Kruskal–Wallis test for comparing K samples subject to unequal pattern of censorship. *Biometrika* 57:579–594.

Breslow, N. E. (1974). Covariance analysis of censored survival data. *Biometrics* 30:89–99.

Breslow, N. E. (1975). Analysis of survival data under the proportional hazards model. *Int. Stat. Rev.* 43:45–58.

Breslow, N. E., and Crowley, J. (1974). A large sample study of the life table and product limit estimates under random censorship. *Ann. Stat.* 2:437–453.

Breslow, N. E., Lubin, J. H., Marek, P., and Langholz, B. (1983). Multiplicative models and cohort analysis. *J. Am. Stat. Assoc.* 78:1–12.

Buckley, T., and James, I. (1979). Linear regression with censored data. *Biometrika* 66:429–436.

Burdette, W. J., and Gehan, E. A. (1970). "Planning and Analysis of Clinical Studies." Thomas, Springfield, Illinois.

Byar, D. P., Huse, R., and Bailar, III. J. C. (1974). An exponential model relating censored survival data and concomitant information for prostatic cancer patients. *J. Nat. Cancer Inst.* 52:321–326.

Byar, D. T. (1983). Analysis of survival data: Cox and Weibull models with covariates. *In* "Statistics in Medical Research: Methods and Issues, with Applications in Clinical Oncology" (V. Mike and K. Stanley, eds.). Wiley, New York.

Chandrasekharan, C., and Hermalin, A. I., eds. (1975). "Measuring the Effects of Family Planning Programs on Fertility." International Union for the Scientific Study of Population, Liege.

Chiang, C. L. (1960a). A stochastic study of the life table and its applications, I. Probability distributions of the biometric functions. *Biometrics* 16:618–635.

Chiang, C. L. (1960b). A stochastic study of the life table and its applications, II. Sample variance of the pobserved expectation of life and other biometric functions. *Hum. Biol.* 32:221–238.

Chiang, C. L. (1961). A stochastic study of the life table and its applications, III. The follow-up study with the consideration of competing risks. *Biometrics* 17:57–78.

Chiang, C. L. (1964). A stochastic model of competing risks of illness and competing risks of death. *In* "Stochastic Models in Medicine and Biology" (J. Gurland, ed.). Univ. of Wisconsin Press, Madison.

Chiang, C. L. (1968). "Introduction to Stochastic Processes in Biostatistics." Wiley, New York.

Chiang, C. L. (1972). On constructing current life tables. *J. Am. Stat. Assoc.* **67**:538–541.

Chiang, C. L. (1978). "The Life Table and Mortality Analysis." The World Health Organization, Geneva.

Chiang, C. L. (1979). Survival and stages of disease. *Math. Biosci.* **43**:159–171.

Chiang, C. L. (1980). "An Introduction to Stochastic Processes and Their Applications." Krieger, Huntington, N. Y.

Chiang, C. L., and van den Berg, B. J. (1982). A fertility table for the analysis of human reproduction. *Math. Biosci.* **62**:237–251.

Clayton, D. G. (1983). Fitting a general family of failure-time distributions using GLIM. *Appl. Stat.* **32**:102–109.

Clayton, D. G., and Kaldor, J. M. (1985). Heterogeneous models as an alternative to proportional hazards in cohort studies. *Gen. Conf. Int. Stat. Inst., Amsterdam, Aug.*

Coale, A. J. (1984). Life table construction on the basis of two enumerations of a closed population. *Population Index* **50**:193–213.

Coale, A. J., and Demeney, P. (1966). "Regional Model Life Tables and Stable Populations." Princeton Univ. Press, Princeton, New Jersey.

Coale, A. J., and Demeny, P. (1983). "Regional Model Life Tables and Stable Populations," 2nd Ed. Academic Press, New York.

Cohen, Jr., A. C. (1951). Estimating parameters of logarithmic-normal distribution by maximum likelihood. *J. Am. Stat. Assoc.* **46**:206–212.

Cohen, Jr., A. C. (1959). Simplified estimators for the normal distribution when samples are singly censored or truncated. *Technometrics* **1**:217–237.

Cohen, Jr., A. C. (1965). Maximum likelihood estimation in the Weibull distribution based on complete and censored samples. *Technometrics* **7**:579–588.

Cohen, Jr., A. C. (1976). Multicensored sampling in the three parameter Weibull distribution. *Technometrics* **17**:347–351.

Cohen, Jr., A. C., and Whitten, B. J. (1980). Estimation in the three parameter lognormal distribution. *J. Am. Stat. Assoc.* **75**:399–404.

Cohen, J., and Cohen, P. (1983). "Applied Multiple Regression/Correlation Analysis for the Behavioral Sciences," 2nd Ed. Erlbaum, Hillsdale, New Jersey.

Connelly, R. R., Cutler, S. J., and Baylis, P. (1966). End result in the cancer of the lung: comparison of male and female patients. *J. Nat. Cancer Inst.* **36**:277–287.

Cornfield, J. (1951). A method of estimating comparative rates form clinical data. Applications to cancer of the lung, breast, and cervix. *J. Natl. Cancer Inst.* **11**:1269–1275.

Cox, D. R. (1953). Some simple approximate tests for Poisson variates. *Biometrika* **40**:354–360.

Cox, D. R. (1959). The analysis of exponentially distributed lifetimes with two types of failure. *J. R. Stat. Soc. Ser. B* **21**:411–421.

Cox, D. R. (1962). "Renewal Theory." Methuen, London.

Cox, D. R. (1972). Regression models and life tables (with discussion). *J. R. Stat. Soc. Ser. B* **34**:187–202.

Cox, D. R. (1975). Partial likelihood. *Biometrika* **62**:269–276.

Cox, D. R. (1978). Some remarks on the role in statistics of graphical methods. *Appl. Stat.* **27**:4–9.

Cox, D. R. (1983). A remark on censoring and surrogate response variables. *J. R. Stat. Soc. Ser. B* **45**:391–393.

Cox. D. R., and Oakes, D. (1984). "Analysis of Survival Data." Chapman & Hall, London.

Cox, D. R., and Snell, E., J. (1968). A general definition of residuals. *J. R. Stat. Soc. Ser. B* **30**:248–275.

Cox, P. R. (1975). "Population Trends," Vols. I and II. Her Majesty's Stationary Office, London.

Crowley, J. (1974). A note on some recent likelihood leading to the log rank test. *Biometrika* **61**:533–538.

Crowley, J., and Hu, M. (1977). Covariance analysis of heart transplant data. *J. Am. Stat. Assoc.* **72**:27–36.

Crowley, J., and Thomas, D. R. (1975). Large sample theory for the log rank test. *Tech. Rep.* (415). Dept of Statistics, Univ of Wisconsin, Madison.

Cutler, S. J., and Ederer, F. (1958). Maximum utilization of the life table method in analyzing survival. *J. Chronic Dis.* **8**:699–712.

Cutler, S. J., and Myers, M. H. (1967). Clinical classification of extent of disease in cancer of the breast. *J. Natl. Cancer Inst.* **39**:193–207.

Cutler, S. J., Ederer, F., Grisworld, M. H., and Greenberg, R. A. (1960a). Survival of patients with uterine cancer, Connecticut, 1935–1954. *J. Natl. Cancer Inst.* **24**:519–539.

Cutler, S. J., Ederer, F., Grisworld, M. H., and Greenberg, R. A. (1960b). Survival of patients with ovarian cancer, Connecticut, 1935–1954. *J. Natl. Cancer Inst.* **24**:541–549.

D'Agostino, R. B. (1971). Linear estimation of the Weibull parameters. *Technometrics* **13**:171–182.

Dandekar, K. (1963). Analysis of birth intervals of a set of Indian women. *Eugenics Quarterly* **10**:73–78.

Dandekar, V. M. (1955). Certain modified forms of binomial and Poisson distributions. *Sankhya* **15**:237–251.

Darby, S. C., and Reissland, J. A. (1981). Low levels of ionizing radiation and cancer—are we understanding the risk? (with discussion). *J. R. Stat. Soc. Ser. A* **144**:298–331

David, H. A. (1970). "Order Statistics." Wiley, New York.

David, H. A., and Moeschberger, M. L. (1978). "The Theory of Competing Risks." Griffin, London.

Davis, D. J. (1952). An analysis of some failure data. *J. Am. Stat. Assoc.* **47**:113–150.

Dempster, A. P., Laird, N. M., and Rubin, D. B. (1977). Maximum likelihood from incomplete data via the EM algorithm (with discussion). *J. R. Stat. Soc. Ser. B* **39**:1–38.

Desu, M., and Narula, S. C. (1977). Reliability estimation under competing causes of failure. *In* "The Theory and Application of Reliability" (C. P. Tsokos, and I. N. Shimi, eds.), vol. 2. Academic Press, New York.

Dixon, W. J. *et al.*, eds. (1981). "BMDP Statistical Software." Univ. of California Press, Berkeley.

Draper, N. R., and Smith, H. (1981). "Applied Regression Analysis," 2nd Ed. Wiley, New York.

Drolette, M. E. (1975). The effect of incomplete follow-up. *Biometrics* **31**:135–144.

Ederer, F., Axtell, L. M., and Cutler, S. J. (1961). The relative survival rate—a statistical methodology. *Natl. Cancer Inst. Monogr.* **6**:101–121.

Efron, B. (1967). The two sample problem with censored data. *Proc. Berkeley Symp., 5th* **4**:831–853.

Efron, B. (1975). The efficiency of logistic regression compared to normal discriminant analysis. *J. Am. Stat. Assoc.* **70**:892–898.

Efron, B. (1977). The efficiency of Cox's likelihood function for censored data. *J. Am. Stat. Assoc.* **72**:555–565.

Efron, B. (1981). Censored data and the bootstrap. *J. Am. Stat. Assoc.* **68**:601–608.

Efron, B., and Hinkley, D. V. (1978). Assessing the accuracy of the maximum likelihood estimator: observed versus expected Fisher information (with discussion). *Biometrika* **65**:457–488.

Elandt-Johnson, R. C. (1973). Age-at-onset distribution in chronic diseases: A life table approach to the analysis of family data. *J. Chronic Dis.* **26**:529–45.

Elandt-Johnson, R. C. (1976). Conditional failure time distributions under competing risk theory with dependent failure times and proportional hazard rates. *Scand. Actuar. J.* **59**:37–51.

Elandt-Johnson, R. C., and Johnson, N. L. (1980). "Survival Models and Data Analysis." Wiley, New York.

Elston, J. S. (1923). Survey of mathematical formulations that have been used to express a law of mortality. *Records* Part I (25):66–86. American Institute of Actuaries.

Engelhardt, M. (1975). On simple estimation of the parameters of the Weibull or extreme value distribution. *Technometrics* **17**:369–374.

Engelhardt, M., and Bain, L. J. (1973). Some complete and censored sammpling results for the Weibull or extreme value distribution. *Technometrics* **15**:541–549.

Engelhardt, M., and Bain, L. J. (1974). Some results on point estimation for the two parameter Weibull or extreme value distribution. *Technometrics* **16**:49–56.

Engelhardt, M., and Bain, L. J. (1977a). Simplified statistical procedures for the Weibull or extreme value distribution. *Technometrics* **19**:323–331.

Engelhardt, M., and Bain, L. J. (1977b). Uniformly most powerful tests on the scale parameter of a gamma distribution with a nuisance shape parameter. *Technometrics* **19**:77–81.

Engelhardt, M., and Bain L. J. (1978a). Construction of optimal unbiased inference procedures for the parameters of the gamma distribution. *Technometrics* **20**:485–489.

Engelhardt, M. and Bain, L. J. (1978b). Tolerance limits and confidence limits on reliability for the two-parameter exponential distribution. *Technometrics* **20**:37–39.

Engelhardt, M., and Bain, L. J. (1979). Prediction limits and two-sample problems with complete or censored Weibull data. *Technometrics* **21**:233–237.

Epstein, B., and Sobel, M. (1953). Life testing. *J. Am. Stat. Assoc.* **48**:486–502.

Espenshade, T. J. (1983). Marriage, divorce, and remarriage from retrospective data: A multiregional approach. *Environ. Planning A* **15**:1633–1652.

Espenshade, T. J. (1985). Marital careers of American women: A cohort life table analysis. *In* "Family Demography: Methods and Their Application" (J. Bongaarts *et al.*, eds.). Oxford Univ. Press, London and New York.

Espenshade, T. J., and Braun, R. (1982). Life course analysis and multistate demography: an application to marriage, divorce, and remarriage. *J. Marriage Family* **44**:1025–1036.

Fabia, J., and Drolette, M. (1970). Life tables up to age 10 for mongols with and without congenital heart defect. *J. Ment. Defic. Res.* **14**:235–42.

Farr. W. (1841). "Fifth Report of the Registrar General," London.

Farr. W. (1874). Effect of the extinction of any single disease on the duration of life. *Suppl. 35th Ann. Rep. Registrar General* **21**:38.

Farewell, V. T., and Prentice, R. L. (1977). A study of distributional shape in life testing. *Technometrics* **19**:69–75.

Farewell, V. T., and Prentice, R. L. (1980). The approximation of partial likelihood with emphasis on case-control studies. *Biometrika* **67**:273–278.

Faulkner, J. E., and McHugh, R. S. (1972). Bias in observable cancer age and life time of mice subject to spontaneous mammary carcinomas. *Biometrics* **28**:489–498.

Feigl, P., and Zelen, M. (1965). Estimation of exponential survival probabilities with concomitant information. *Biometrics* **21**:826–838.

Feinleib, M. (1960). A method of analyzing lognormally distributed survival data with incomplete follow-up. *J. Am. Stat. Assoc.* **55**:534–545.

Feinstein, A. R. (1977). "Clinical Biostatistics." Mosby, St. Louis.

Feller, W. (1968). "An Introduction to Probability Theory and Its Applications," Vol. I, 3rd Ed. Wiley, New York.

Fertig, K. W., and Mann, N. R. (1980). Life test sampling plans for two-parameter Weibull populations. *Technometrics* **22**:165–177.

Fix, E., and Neyman, J. (1951). A simple stochastic model of recovery, relapse, death and loss of patients. *Hum. Biol.* **23**:205–241.

Fleiss, J. L. (1983). "Statistical Methods for Rates and Proportions," 2nd Ed. Wiley, New York.

Fleming, T. R. (1978a). Nonparametric estimation for nonhomogeneous Markov processes in the problem of competing risks. *Ann. Stat.* **6**:1057–1070.

Fleming, T. R. (1978b). Asymptotic distribution results in competing risks estimation. *Ann. Stat.* **6**:1071–1–79.

Fleming, T. R., O'Fallon, J. R., O'Brien, P. C., and Harrington, D. P. (1980). Modified Kolmogorov–Smirnov test procedures with applications to arbitrarily right censored data. *Biometrics* **36**:607–626.

Freireich, E. O.*et al.* (1963). The effect of 6-mercaptopurine on the duration of steroid induced remission in acute leukemia. *Blood* **21**:699–716.

Freireich, E. J. *et al.* (1974). New prognostic factors affecting response and survival in adult acute leukemia. *Trans. Assoc. Am. Physicians* **87**:298–305.

Gaddum, J. H. (1945a). Log normal distributions. *Nature (London)* **156**:463.

Gaddum, J. H. (1945b). Log normal distributions. *Nature (London)* **156**:747.

Gail, M. H. (1975). A review and critique of some models used in competing risk analysis. *Biometrics* **31**:209–222.

Gail, M. H., and Gastwirth, J. L. (1978). A scale-free goodness of fit test for the exponential distribution based on the Gini statistics. *J. R. Stat. Soc. Ser. B* **40**:350–357.

Gail, M. H., and Ware, J. (1979). Comparing observed life table data with a known survival curve in the presence of random censorship. *Biometrics* **35**:385–391.

Gajjar, A. V., and Khatri, C. G. (1969). Progressively censored samples from log-normal and logistic distributions. *Technometrics* **11**:793–803.

Galambos, J. (1978). "The Asymptotic Theory of Extreme Order Statistics." Wiley, New York.

Galambos. J., and Kotz, S. (1978). Characterizations of probability distributions. "Lecture Notes in Mathematics," Vol. 675. Springer-Verlag, Berlin and New York.

Gehan, E. A. (1965a). A generalized Wilcoxon test for comparing arbitrarily singly censored samples. *Biometrika* **52**:203–223.

Gehan, E. A. (1965b). A generalized two-sample Wilcoxon test for doubly-censored data. *Biometrika* **52**:650–653.

Gehan, E. A. (1969). Estimating survival functions from the life table. *J. Chronic Dis.* **21**:629–644.

Gehan, E. A., and Siddiqui, M. M. (1973). Simple regression methods for survival time studies. *J. Am. Stat. Assoc.* **68**:848–856.

Gehan. E. A., and Thomas, D. G. (1969). The performance of some two-sample tests in small samples with and without censoring. *Biometrika* **56**:127–132.

Gershenson, H. (1961). "Measurement of Mortality." Society of Actuaries, Chicago.

Gibbons, J. D. (1976). "Nonparametric Methods for Quantitative Analysis." Holt, New York.

Glasser, M. (1965). Regression analysis with dependent variable censored. *Biometrics* **21**:300–307.

Glasser, M. (1967). Exponential survival with covariance. *J. Am. Stat. Assoc.* **62**:561–568.

Gompertz, B. (1825). On the nature of the function expressive of the law of human mortality. *Philos. Trans. R. Soc.* **115**:513–593.

Graunt, J. (1666). Natural and Political Observations...upon the Bills of Mortality. London.

Greenberg, R., Bayard, S., and Byar, D. (1974). Selecting concomitant variables using a likelihood ratio step-down procedure and a method of testing goodness of fit of an exponential survival model. *Biometrics* **30**:601–608.

Greenwood, J. A., and Durand, D. (1960). Aids for fitting the gamma distribution by maximum likelihood. *Technometrics* **2**:55–65.

Greenwood, M. (1922). On the value of life-tables in statistical research (with discussion). *J. R. Stat. Soc.* **85**:537–560.

Greenwood, M. (1926). The natural duration of cancer. "Reports on Public Health and Medical Subjects," Vol. 33. Her Majesty's Stationery Office, London.

Greville, T. N. E. (1943). Short methods of constructing abridged life tables. *Rec. Am. Inst. Actuar.* **32**:29–43.

Greville, T. N. E. (1946). "United States Life Tables and Actuarial Tables, 1939–1941." U. S. Bureau of the Census, Washington, D. C.

Greville, T. N. E. (1948). Mortality tables analyzed by cause of death. *Rec. Am. Inst. Actuar.* **37**:283–294.

Griffith, J. D., Suchindran, C. M., and Koo, H. P. (1985). Cohort change in the marital life course: An analysis of marital life histories using multistate life table techniques. Working Paper Number 37. Carolina Population Center, Chapel Hill.

Grizzle, J. E. (1967). Continuity correction in the χ^2-test for 2×2 tables. *Am. Stat.* **21**:28–32.

Grizzle, J. E., Starmer, C. F., and Koch, G. G. (1969). Analysis of categorical data by linear models. *Biometrics* **25**:489–504.

Gross, A. J., and Clark, V. A. (1975). "Survival Distributions: Reliability Applications in the Biomedical Sciences." Wiley, New York.

Gumbel, E. J. (1958). "Statistics of Extreme." Columbia Univ. Press, New York.

Gupta, S. S. (1960). Order statistics from the gamma distribution. *Technometrics* **2**:243–262.

Haberman, S. (1983). Decrement tables and the measurement of morbidity I. *J. Inst. Actuar.* **110**:361–381.

Haberman, S. (1984). Decrement tables and the measurement of morbidity II. *J. Inst. Actuar.* **111**:73–86.

Hager, H. W., and Bain, L. J. (1970). Inferential procedures for the generalized gamma distribution. *J. Am. Stat. Assoc.* **65**:1601–1609.

Hall, W. J., and Wellner, J. A. (1980). Confidence bands for a survival curve from censored data. *Biometrika* **67**:133–143.

Halley, E. (1693). An estimate of the degrees of the mortality of mankind, drawn from curious tables of the births and funerals at the city of Breslau. *Philos. Trans. R. Soc.* **17**:596–610.

Hannan, M. T., and Carrol, G. R. (1981). Dynamics of formal political structure: An event-history analysis. *Am. Sociol. Rev.* **46**:19–35.

Hannan, M. T., and Tuma, N. B. (1979). Methods for temporal analysis. *Annu. Rev. Sociol.* **5**:303–328.

Hannan, M. T., Tuma, N. B., and Groeneveld, L. P. (1977). Income and marital event: Evidence from an income maintenance experiment. *Am. J. Sociol.* **82**:1186–1211.

Hannan, M. T., Tuma, N. B., and Groeneveld, L. P. (1978). Income and independence effect on marital resolution: Results from the Seattle and Denver income maintenance experiments. *Am. J. Sociol.* **84**:611–633.

Harrell, F. (1980a). The PHGLM procedure. *SAS Inst. Tech. Rep.* S–109. SAS Institute, Raleigh, North Carolina.

Harrell, F. (1980b). The LOGIST procedure. *SAS Inst. Tech. Rep.* S–110. SAS Institute, Raleigh, North Carolina.

Harter, H. L., and Moore, A. H. (1965). Maximum likelihood estimation of the parameters of gamma and Weibull populations from complete and from censored samples. *Technometrics* **7**:639–643.

Harter, H. L., and Moore, A. H. (1966). Local maximum likelihood estimation of the parameters of three-parameter log-normal population from complete and censored samples. *J. Am. Stat. Assoc.* **61**:842–851.

Harter. H. L., and Moore, A. H. (1968). Maximum likelihood estimation, from doubly censored samples, of the parameters of the first asymptotic distribution of extreme values. *J. Am. Stat. Assoc.* **63**:889–901.

Hayward, M. D., Grady, W. R., McLaughlin, S. D., and Armstrong, T. L. (1985). Occupational differences in the labor force behavior of older men and women for the United States, 1972: An application of an increment–decrement working life table. *Annu. Meet. Popul. Assoc. Am., Boston,* Abstract, Population Index 51.

Heckman, J. J. (1979). Sample selection bias as a specification error. *Econometrica* **47**:153–161.

Heckman, J. J., and Singer, B. (1982). Population heterogeneity in demographic models. *In* "Multidimensional Mathematical Demography" (K. Land, and A. Rogers, eds.). Academic Press, New York.

Heckman, J. J., and Singer, B. (1984). Econometric duration analysis. *J. Econometr.* **24**:63–132.

Heligman, L., and Pollard, J. H. (1980). The age pattern of mortality. *J. Inst. Actuar.* **107**:49–80.

Hobcraft, J., and Rodriguez, R. (1980). Methodological issues in life table analysis of birth histories. *Seminar on the Analysis of Maternity Histories.* London School of Hygiene and Tropical Medicine.

Hoel, D. G. (1972). A representation of mortality by competing risks. *Biometrics* **22**:99–108.

Hoem, J. M. (1969). Purged and partial Markov chains. *Scand. Actuar. J.,* 147–155.

Hoem, J. M. (1970). Probabilistic fertility models of the life table type. *Theor. Popul. Biol.* **1**:12–38.

Hoem, J. M. (1971). On the interpretation of certain vital rates as averages of underlying forces of transition. *Theor. Popul. Biol.* **2**:454–458.

Hoem, J. M. (1975). The construction of increment–decrement life tables: A comment on articles by R. Schoen and V. Nelson. *Demography* **12**:661–665.

Hoem, J. M. (1977). A Markov chain model of working life tables. *Scand. Actuar. J.,* 1–20.

Hoem, J. M., and Fong, M. S. (1976). A Markov chain model of working life tables. Working Paper No. 2. Laboratory of Actuarial Mathematics, Univ. of Copenhagen.

Hoem, J. M., and Jensen, U. F. (1982). Multistate life table methodology: A probabilistic critique. *In* "Multidimensional Mathematical Demography" (K. C. Land, and A. Rogers, eds.). Academic Press, New York.

Hofferth, S. L. (1985a). Updating children's life course. *J. Marriage and the Family* **47**:93–115.

Hofferth, S. L. (1985b). Recent trends in the living arrangements of children: A cohort life table analysis. *In* "Family Demography Methods and their Application." (Bongaarts *et al.*, eds.). Oxford Univ Press, Oxford.

Holford, T. R. (1976). Life tables with concomitant information. *Biometrics* **32**:587–597.

Holford, T. R. (1980). The analysis of rates and survivorship using log-linear models. *Biometrics* **36**:299–306.

Holland, B. (1983). Breastfeeding and infant mortality: A hazards model analysis of the case of Malaysia. Unpublished Ph.D. dissertation, Princeton University.

Horiuchi, S., and Coale, A. J. (1982). A simple equation for estimating the expectation of life at old ages. *Popul. Stud.* **36**:317–326.

Hougaard, P. (1984a). Life table methods for heterogenous populations: Distributions describing the heterogeneity. *Biometrika* **71**:75–83.

Hougaard, P. (1984b). Frailty models derived from the stable distributions. Institute of Mathematical Statistics, University of Copenhagen.

Hyrenius, H., and Quist, J. (1970). Life table technique for working ages. *Demography* **7**:393–399.

Irwin, A. C. (1976). Life tables as 'predictors' of average longevity. *Can. Med. Assoc.* **114**:539–541.

Irwin, J. O. (1949). Standard error of an estimate of expectation of life. *J. Hyg.* **47**:188–189.

Jain, A. K., and Sivin, I. (1977). Life-table analysis of IUDs: Problems and recommendations. *Stud. Family Planning* **8**:26–47.

Johansen S. (1978). The product limit estimate as a maximum likelihood estimate. *Scand. J. Stat.* **5**:195–199.

Johnson, N. L. (1977). Approximate relationships among some estimators of mortality probabilities. *Biometrics* **33**:542–545.

Johnson, N. L., and Kotz, S. (1969). "Discrete Distributions." Houghton Mifflin, Boston.

Johnson, N. L., and Kotz, S. (1970). "Continuous Univariate Distributions," Vols. 1 and 2. Houghton Mifflin, Boston.

Johnson, N. L., and Kotz, S. (1972). "Distributions in Statistics: Continuous Multivariate Distributions." Wiley, New York.

Johnson, N. L., and Kotz, S. (1975). A vector-valved multivariate hazard rate. *J. Multivar. Anal.* 5:53–66.

Jordan, C. W. (1967). "Life Contingencies." Soc. of Actuaries, Chicago.

Kahn, H. A. (1966). The Dorn study of smoking and mortality among U. S. Veterans: Report on eight and one-half years of observation. *In* "Epidemiological Approaches to the Study of Cancer and other Chronic Diseases" (W. Haenzel, ed.). *Natl. Cancer Inst. Monogr.* (19).

Kalbfleisch, J. D. (1974). Some efficiency calculations for survival distributions. *Biometrika* 61:31–38.

Kalbfleisch, J. D., and Lawless, J. F. (1985). The analysis of panel data under a Markov assumption. *J. Am. Stat. Assoc.* 80:863–871.

Kalbfleisch, J. D., and McIntosh, A. (1977). Efficiency in survival distributions with time-dependent covariates. *Biometrika* 64:47–50.

Kalbfleisch, J. D., and Prentice, R. L. (1973). Marginal likelihood based on Cox's regression and life model. *Biometrika* 60:267–278.

Kalbfleisch, J. D., and Prentice, R. L. (1980). "The Statistical Analysis of Failure Time Data." Wiley, New York.

Kalbfleisch, J. D., and Sprott, D. A. (1970). Application of likelihood methods to models involving a large number of parameters (with discussion) *J. R. Stat. Soc. Ser. B* 32:175–208.

Kalbfleisch, J. D., and Sprott, D. A. (1974). Marginal and conditional likelihood. *Sankhya A* 35:311–328.

Kaplan, E. L., and Meier, P. (1958). Nonparametric estimation from incomplete observations. *J. Am. Stat. Assoc.* 53:457–481.

Kay, R. (1977). Proportional hazard regression models and the analysis of censored survival data. *Appl. Stat.* 26:227–237.

Kay, R. (1979). Some further asymptotic efficiency calculations for survival data regression models. *Biometrika* 66:91–96.

Keyfitz, N. (1966). A life table that agrees with the data. *J. Am. Stat. Assoc.* 61:305–312.

Keyfitz, N. (1968). "Introduction to the Mathematics of Population. Addison-Wesley, Reading, Massachusetts.

Keyfitz, N. (1970). Finding probabilities from observed rates, or how to make a life table. *Am. Stat.* 24:28–33.

Keyfitz, N. (1977). "Applied Mathematical Demography." Wiley, New York.

Keyfitz, N. (1979). Multidimensionality in population analysis. *In* "Sociology Methodology 1980." (K. Schuessler, ed.). Jossey-Bass, San Francisco.

Keyfitz, N. (1980). Multistate demography and its data: A comment. *Environ. Planning A* 12:615–622.

Keyfitz, N. (1981). Choice of function for mortality analysis. *Theor. Popul. Biol.* 21:329–352.

Keyfitz, N. (1984). Heterogeneity and selection in population analysis. Research Paper no. 10. Statistics Canada, Ottawa, Canada.

Keyfitz, N., and Frauenthal, J. (1975). An improved life table method. *Biometrics* 31:889–899.

Keyfitz, N., and Littman, G. (1980). Mortality in a heterogeneous population. *Popul. Stud.* 33:333–343.

Keyfitz, N., Preston, S., and Schoen, R. (1972). Inferring probabilities from rates: Extensions to multiple decrements. *Scand. Act. J.*, 1–13.

Kimball, A. W. (1958). Disease incidence estimation in population subject to multiple causes of death. *Int. Stat. Inst. Bull.* 36:193–204.

Kimball, A. W. (1969). Models for the estimation of competing risk from grouped data. *Biometrics* 25:329–337.

Koch, G. G. Johnson, E. D., and Tolly, H. D. (1972). A linear model approach to the analysis of survival and extent of disease in multidimensional contingency tables. *J. Am. Stat. Assoc.* **67**:783–796.

Koesoebjono, S. (1979). Marital status life tables of female population in the Netherlands, 1978: An application of multidimensional demography. Working Paper 20. Voorburg, The Netherlands: NIDI

Koo, H. P., Suchindran, C. M., and Griffith, J. D. (1984). The effects of children on divorce and remarriage: A multivariate analysis of lifetable probabilities. *Popul. Stud.* **38**:451–471.

Krall, J. M., Uthoff, V. A., and Harley, J. B. (1975). A step-up procedure for selecting variables associated with survival. *Biometrics* **31**:49–57.

Krishnamoorthy, S. (1979). Classical approach to increment–decrement life tables: An application to the study of the marital status of United States females, 1970. *Math. Biosci* **44**:139–154.

Krishnamoorthy, S. (1982). Marital status life tables for Australian women, 1971. *Genus* **38**:97–117.

Krishnan, P. (1971). Divorce tables for females in the United States: 1960. *J. Marriage and the Family* **33**:318–320.

Kruskal, W. H., and Wallis, W. A. (1952). Use of ranks in one-criterion analysis of variance. *J. Am. Stat. Assoc.* **47**:583–621.

Kuzma, J. W. (1967). A comparison of two life table methods. *Biometrics* **23**:51–64.

Lagakos, S. W. (1979). General right censoring and its impact on the analysis of survival data. *Biometrics* **35**:139–156.

Lagakos, S. W. (1981). The graphical evaluation of explanatory variables in proportional hazard regression models. *Biometrika* **68**:93–98.

Lagakos, S. W., Sommer, C. J., and Zelen, M. (1978). Semi-Markov models for partially censored data. *Biometrika* **65**:311–317.

Laird, N., and Olivier, D. (1981). Covariance analysis of survival data using log-linear analysis techniques. *J. Am. Stat. Assoc.* **76**:231–240.

Lancaster, T. (1985). Generalized residuals and heterogeneous duration models. *J. Econometr.* **28**:155–169.

Land, K., and Hough, Jr., G. C. (1985). Voting status life tables for the U.S., 1968–1980. *Annu. Meet. Popul. Assoc. Am., Boston.*

Land, K., and Rogers, A., eds. (1982). "Multidimensional Mathematical Demography." Academic Press, New York.

Langberg, N., Proschan, F., and Quinzi, A. J. (1978). Converting dependent models into independent ones, preserving essential features. *Ann. Probability* **6**:174–181.

Langberg, N., Proschan, F., and Quinzi, A. J. (1981). Estimating dependent life lengths, with application to the theory of competing risks. *Ann. Stat.* **9**:157–167.

Langlands, A. O., Pocock, S. J., Kerr, G. R., and Gore, S. M. (1979). Long-term survival of patients with breast cancer: A study of the curability of the disease. *Br. Med. J.* **2**:1247–12251.

Lawless, J. F. (1971). A prediction problem concerning samples from the exponential distribution with application in life testing. *Technometrics* **4**:725–730.

Lawless, J. F. (1973). On the estimation of safe life when the underlying distribution is Weibull. *Technometrics* **15**:857–865.

Lawless, J. F. (1975). Construction of tolerance bounds for the extreme value and Weibull distributions. *Technometrics* **17**:255–261.

Lawless, J. F. (1978). Confidence interval estimation for the Weibull and extreme value distributions. *Technometrics* **20**:355–364.

Lawless, J. F. (1980). Inference in the generalized gamma and log gamma distributions. *Technometrics* **22**:409–419.

Lawless, J. F. (1982). "Statistical Models and Methods for Lifetime Data." Wiley, New York.

Lawless, J. F., and Mann, N. R. (1976). Tests for homogeneity for extreme value scale parameters. *Commun. Stat. A* **5**:389–405.

Lawless, J. F., and Singhal, K. (1978). Efficient screening of non-normal regression models. *Biometrics* **34**: 318–327.

Lawless, J. F., and Singhal, K. (1980). Analysis of data from life test experiments under an exponential model. *Nav. Res. Log Q.* **27**:323–334.

Ledent, J. (1980). Multistate life tables: Movement versus transition perspectives. *Environ. Planning A* **12**:533–562.

Ledent, J., and Rees, P. H. (1986). Life tables. *In* "Migration and Settlement: A Multiregional Comparative Study" (A. Rogers, and F. J. Willekens, eds.). Reidel, Dordrecht.

Ledermann, S. (1969). Nouvelles tables-types de mortalité. Travaux et Documents, Cahiers, No. 53. Institut National d'Etudes Dámographiques, Paris.

Lee, E. T. (1980). "Statistical Methods for Survival Data Analysis." Lifetime Learning Publications, Belmont, California.

Lee, E. T., Desu, M. M., and Gehan, E. A. (1975). A Monte Carlo study of the power of some two-sample tests. *Biometrika* **62**:425–432.

Lee, L. (1980). Testing adequacy of the Weibull and log linear rate models for a Poisson process. *Technometrics* **22**:195–200.

Lee, L., and Lee, S. K. (1978). Some results on inference for the Weibull process. *Technometrics* **20**:41–45.

Lehrer, E. (1984). The impact of child mortality on spacing by parity: A Cox regression analysis. *Demography* **21**:323–338.

Lemon, G. (1975). Maximum likelihood estimation for the three-parameter Weibull distribution, based on censored samples. *Technometrics* **17**:247–254.

Lew, E. A., and Seltzer, F. (1970). Uses of the life table in public health. *Milbank Memor. Fund Q.* **48**(Suppl.):15–37.

Liard, N., and Olivier, D. (1981). Covariance analysis of censored survival data using log-linear analysis techniques. *J. Am. Stat. Assoc.* **76**:231–240.

Liaw, K. L. (1978). Dynamic properties of the 1966–1971 Canadian spatial population system. *Environ. Planning A* **10**:289–298.

Liaw, K. L. (1980). Multistate dynamics: The convergence of an age-by-region population system. *Environ. Planning A* **12**:589–613.

Liaw, K. L. (1986). Spatial population dynamics. *In* "Migration and Settlement: A Multiregional Comparative Study" (A. Rogers and F. Willekens, eds.). Reidel, Dordrecht.

Lieblein, J., and Zelen, M. (1956). Statistical investigation of the fatigue life of deep groove ball bearings. *J. Res. Natl. Bur. Stand.* **57**:273–316.

Lininger, L., *et al.* (1979). Comparison of four tests for equality of survival curves in the presence of stratification and censoring. *Biometrika* **66**:419–428.

Littell, A. S. (1952). Estimation of the T-year survival rate for follow-up studies over a limited period of time. *Hum. Biol.* **24**:87–116.

Littell, R. C., McClave, J. T., and Offen, W. W. (1979). Goodness of fit tests for two-parameter Weibull distribution. *Commun. Stat. B* **8**:257–269.

Louis, T. A. (1977). Sequential allocation in clinical trials comparing two exponential curves. *Biometrics* **33**:627–634.

Makeham, W. M. (1860). On the law of mortality and the construction of annuity tables. *J. Inst. Actuar.* **8**:301–310.

Mann, N. R. (1968). Point and interval estimation procedures for the two-parameter Weibull and extreme value distributions. *Technometrics* **10**:231–256.

Mann, N. R., and Fertig, K. W. (1973). Tables for obtaining confidence bounds and tolerance bounds based on best linear invariant estimates of parameters of the extreme value distribution. *Technometrics* **15**:87–101.

Mann, N. R., Schafer, R. E., and Singpurwalla, N. D. (1974). "Methods for Statistical Analysis of Reliability and Lifetime Data." Wiley, New York.

Mantel, N. (1963). Chi-square tests with one degree of freedom: Extensions of the Mantel–Haenszel procedure. *J. Am. Stat. Assoc.* **58**:690–700.

Mantel, N. (1966). Evaluation of survival data and two new rank order statistics arising in its consideration. *Cancer Chemother. Rep.* **50**:163–170.

Mantel, N. (1967). Ranking procedures for arbitrarily restricted observations. *Biometrics* **23**:65–78.

Mantel, N., and Haenszel, W. (1959). Statistical aspects of the anslysis of data from retrospective studies of disease. *J. Natl. Cancer Inst.* **22**:719–748.

Mantel, N., and Myers, M. (1971). Problems of convergence of maximum likelihood iterative procedures in multiparameter situations. *J. Am. Stat. Assoc.* **66**:484–491.

Manton, K. G., Poss, S. S., and Wing, S. (1979). The black/white mortality crossover: Investigation from the perspectives of the components of aging. *Gerontologist* **19**:291–299.

Manton, K. G., and Stallard, E. (1981a). Heterogeneity and its effect on mortality measurement. *In* "Methodologies for the Collection and Analysis of Mortality Data" (J. Vallen, J. H. Pollard, and L. Heligman, eds.). International Union for the Scientific Study of Population, Liege.

Manton, K. G., and Stallard, E. (1981b). Methods for the analysis of mortality risks across heterogenous small populations: Examination of space–time gradients in cancer mortality in North Carolina counties 1970–1975. *Demography* **18**:217–230.

Manton, K. G., and Stallard, E. (1981c). Methods for evaluating the heterogeneity of aging processes in human populations using vital statistics data: Explaining the black/white mortality crossover by a model of mortality selection. *Hum. Biol.* **53**:47–67.

Manton, K. G., and Stallard, E. (1982). The use of mortality time series data to produce hypothetical morbidity distributions and project mortality trends. *Demography* **19**:223–240.

Manton, K. G., and Stallard, E. (1984). "Recent Trends in Mortality Analysis." Academic Press, Orlando, Florida.

Manton, K. G., Tolley, H. D., and Poss, S. S. (1976). Life table techniques for multiple-cause mortality. *Demography* **13**:541–564.

Manton, K. G., Woodbury, M. A., and Stallard, E. (1979). Analysis of the components of CHD risk in the Farmingham study: New multivariate procedures for the analysis of chronic disease development. *Comput. Biomed. Res.* **12**:109–123.

Manton, K. G., Stallard, E., and Vaupel, J. W. (1981a). Methods for comparing the mortality experience of heterogeneous populations. *Demography* **18**:389–410.

Manton, K. G., Woodbury, M. A., and Stallard, E. (1981b). A variance component approach to categorical data models with heterogeneous cell populations: Analysis of spatial gradients in lung cancer mortality rates in North Carolina counties. *Biometrics* **37**:259–269.

Martin, L. G., Trussel, J., Salvail, F. R., and Shah, N. (1983). Covariates of child mortality in the Philippines, Indonesia, and Pakistan: An analysis based on hazards models. *Popul. Stud.* **37**:417–432.

Meier, P. (1975). Estimation of a distribution function from incomplete observations. *In* "Perspectives in Probability and Statistics" (J. Gani, ed.). Academic Press, London.

Menken, J. A. (1975). Biometric models of fertility. *Social Forces* **54**:52–64.

Menken, J. A., and Sheps, M. C. (1972). The sampling frame as a determinant of observed distributions of duration variables. *In* "Population Dynamics" (T. N. E. Greville, ed.). Academic Press, New York.

Menken, J., Trussel, J., Stempel, D., and Babakol, O. (1981). Proportional hazards life table models: An illustrative analysis of socio-demographic influences on marriage dissolution in the United States. *Demography* **18**:181–200.

Michael, R. T., and Tuma, N. B. (1985). Entry into marriage and parenthood by young men and women: The influence of family background, *Demography* **22**:515–544.

Miller, M. D. (1946). "Elements of Graduation." Actuarial Society of America, Chicago.

Miller, R. B., and Hickman, J. C. (1983). A comparison of several life table estimation methods. *Scand. Actuar. J.* **66**:77–86.

Miller, Jr., R. G. (1976). Least-squares regression with censored data. *Biometrika* **63**:449–464.

Miller, Jr., R. G. (1981). "Survival Analysis." Wiley, New York.

Miller, Jr., R. G., and Halpern, J. (1982). Regression with censored data. *Biometrika* **69**:521–531.

Mitra, S. (1973). On the efficiency of the estimates of life table functions. *Demography* **10**:421–426.

Mitra, S. (1984). Estimating the expectation of life at older ages. *Popul. Stud.* **38**:313–319.

Mode, C. J. (1982). Increment–decrement life tables and semi-Markovian processes from a sample path perspective. *In* "Multidimensional Mathematical Demography" (K. C. Land and A. Rogers, eds.). Academic Press, New York.

Mode, C. J. (1985). "Stochastic Processes in Demography and Their Computer Implementation." Springer-Verlag, Berlin.

Mode, C. J., and Busby, R. C. (1982). An eight parameter model of human mortality—the single decrement case. *Bull. Math. Biol.* **44**:647–659.

Moeschberger, M. L. (1974). Life tests under competing causes of failure. *Technometrics* **16**:39–47.

Myers, M. H., Axtell, L. M., and Zellen, M. (1966). The use of prognostic factors in predicting survival for breast cancer patients. *J. Chronic Dis.* **19**:923–933.

Myers, M., Hankey, B. G., and Mantel, N. (1973). A logistic-exponential model for use with response-time data involving regressor variables. *Biometrics* **29**:257–269.

Nádas, A. (1970). On proportional hazard functions. *Technometrics* **12**:413–416.

Nádas, A. (1971). The distribution of the identified minimum of a normal pair determines the distribution of the pair. *Technometrics* **13**:201–202.

Namboodiri, N. K. (1984). "Matrix Algebra: An Introduction." Sage, Beverly Hills.

Namboodiri, N. K. (1972). Experimental designs in which each subject is used repeatedly. *Psychol. Bull.* **77**:54–64.

Namboodiri, N. K., and West, K. (1978). On the danger of "overfitting" the data in the application of the GSK method to the analysis of categorical data. *Proc. Social Stat. Sect. Am. Stat. Assoc., Washington, D.C.*

Namboodiri, N. K., Carter, L. F., and Blalock, Jr., H. M. (1975). "Applied Multivariate Analysis and Experimental Designs." McGraw-Hill, New York.

Nelson, W. B. (1970). Hazard plotting methods for analysis of life data with different failure modes. *J. Qual. Technol.* **2**:126–149.

Nelson, W. B. (1972). Theory and application of hazard plotting for censored failure data. *Technometrics* **14**:945–965.

Nelson, W. B. (1982). "Applied Life Data Analysis." Wiley, New York.

Nelson, W. B., and Hahn, G. J. (1972). Linear estimation of regression relationship from censored data, part I: Simple methods and their applications. *Technometrics* **14**:247–269.

Nelson, W. B., and Hahn, G. J. (1973). Linear estimation of regression relationship from censored data, part II: Best unbiased estimation and theory (with discussion). *Technometrics* **15**:133–150.

Nelson, W. B., and Schmee, J. (1979). Inference for (log) normal life distributions from small singly censored samples and blue's. *Technometrics* **21**:43–54.

Neter, J., and Wasserman, W. (1974). "Applied Linear Statistical Models." Irwin, Homewood, Illinois.

New Zealand Census and Statistics Department (1955). Tables of working life. *Monthly Abstr. Stat. Suppl.* **November**.

Nour, E.-S. (1984). Mortality of a heterogeneous cohort; descriptions and implications. *Biometr. J.* **26**:931–940.

Nour, E.-S., and Suchindran, C. M. (1983). A general formulation of the life table. *Math. Biosci.* **63**:241–252.

Nour, E.-S., and Suchindran, C. M. (1984a). The contribution of multistate life tables: Comments on the article by Willekens *et al. Popul. Stud.* **38**:325–328.

Nour, E.-S., and Suchindran, C. M. (1984b). Multistate mortality by cause of death: A life table analysis. *J. R. Stat. Soc. Ser. A* **147**:582–597.

Nour, E.-S., and Suchindran, C. M. (1985). Multistate life tables: Theory and applications. *In* "Biostatistics: Statistics in Biomedical, Public Health, and Environmental Sciences" (P. K. Sen, ed.). North-Holland Publ., Amsterdam.

Oakes, D. (1977). The asymptotic information in censored survival data. *Biometrika* **64**:441–448.

Oakes, D. (1981). Survival times: Aspects of partial likelihood (with discussion). *Int. Stat. Rev.* **49**:199–233.

Oakes, D. (1982a). A concordance test for independence in the presence of censoring. *Biometrics* **38**:451–455.

Oakes, D. (1982b). A model for association in bivariate survival data. *J. R. Stat. Soc. Ser. B* **44**:414–422.

Oechsli, F. W. (1975). A population model based on a life table that includes marriage and parity. *Theor. Popul. Biol.* **2**:229–245.

Parker, R. L., Dry, T. J., Willius, F. A., and Gage, R. P. (1946). Life expectancy in Angina Pectoris. *J. Am. Med. Assoc.* **131**:95–100.

Perrin, E. B., and Sheps, M. C. (1964). Human reproduction: A stochastic process. *Biometrics* **20**:28–45.

Peterson, A. V. (1977). Expressing the Kaplan–Meier estimator as a function of empirical subsurvival functions. *J. Am. Stat. Assoc.* **72**:854–858.

Petit, A. N., and Stephens, M. A. (1976). Modified Cramer–von Mises statistics for censored data. *Biometrika* **63**:291–298.

Peto, R. (1972). Contribution to discussion of paper by D. R. Cox. *J. R. Stat. Soc. Ser. B* **34**:205–207.

Peto, R., and Lee, P. (1973). Weibull distributions for continuous carcinogenesis experiments. *Biometrics* **29**:457–470.

Peto, R., and Peto, J. (1972). Asymptotically efficient rank invariant procedures (with discussion). *J. R. Stat. Soc. Ser. A* **135**:185–206.

Peto, R., and Pike, M. C. (1973). Conservatism of the approximation $(0 - E)^2/E$ in the log rank test for survival data or tumor incidence data. *Biometrics* **29**:579–584.

Peto, R., Lee, P. N., and Paige, W. S. (1972). Statistical analysis of the bioassay of continuous carcinogens. *Brit. J. Cancer* **26**:258–261.

Peto, R., Pike, M. C., Armitage, P., Breslow, N. E., Cox, D. R., Howard, S. V., Mantel, N., McPherson, K., Peto, J., and Smith, P. G. (1977). Design and analysis of randomized and clinical trials requiring prolonged observations of each patient. Part II. Analysis and examples. *Br. J. Cancer* **35**:1–29.

Philipov, D., and Rogers, A. (1981). Multistate population projections. *In* "Advances in Multiregional Demography" (A. Rogers, ed.). International Institute for Applied Systems Analysis, Luxenberg, Austria.

Philipov, D., and Rogers, A. (1982). Multiregional population projections by place of previous residence. *In* "Multidimensional Mathematical Demography" (K. Land and A. Rogers, eds.). Academic Press, New York.

Pierce, D. A., Stewart, W. H., and Kopecky, K. J. (1979). Distribution-free analysis of grouped survival data. *Biometrics* **35**:785–793.

Pierce, M., *et al.* (1969). Epidemiological factors and survival experience in 1770 children with acute leukemia. *Cancer* **23**:1296–1304.

Pike, M. C. (1966). A suggested method of analysis of a certain class of experiments in carcinogenesis. *Biometrics* **22**:142–161.

Potter, R. G. (1963). Birth intervals—Structure and change. *Popul. Stud.* **17**:155–166.

Potter, R. G. (1967). The multiple decrement life table as an approach to the measurement of use effectiveness and demographic effectiveness of contraception. Contributed papers, Sydney Conference of the International Union for the Study of Population.

Potter, R. G. (1969). Use-effectiveness of intrauterine contraception as a problem in competing risks. *In* "Family Planning in Taiwan" (R. Freedman and J. Y. Takeshita, eds.). Princeton Univ. Press, Princeton; N.J.

Prentice, R. L. (1973). Exponential survival with censoring and explanatory variables. *Biometrika* **60**:279–288.

Prentice, R. L. (1974). A log gamma model and its maximum likelihood estimation. *Biometrika* **61**:539–544.

Prentice, R. L. (1975). Discrimination among some parametric models. *Biometrika* **62**:607–614.

Prentice, R. L. (1978). Linear rank tests with right-censored data. *Biometrika* **65**:167–179.

Prentice, R. L., and Gloeckler, L. A. (1978). Regression analysis of grouped survival data with application to breast cancer data. *Biometrics* **34**:57–67.

Prentice, R. L., and Kalbfleisch, J. D. (1979). Hazard rate models with covariates. *Biometrics* **35**:25–39.

Prentice, R. L., and Marek, P. (1979). A qualitative discrepancy between censored data rank tests. *Biometrics* **35**:861–867.

Prentice, R. L., Kalbfleisch, J. D., Peterson, A. V., Flournoy, N., Farewell, V. T., and Breslow, N. E. (1978). The analysis of failure times in the presence of competing risks. *Biometrics* **34**:541–554.

Pressat, R. (1972). "Demographic Analysis" (trans. J. Matras) Aldine, Chicago, Illinois.

Preston, S. (1976). "Mortality Patterns in National Populations with Special Reference to Recorded Causes of Death." Academic Press, New York.

Preston, S. H., Keyfitz, N., and Schoen, R. (1972). "Causes of Death Life Tables for National Populations." Seminar Press, New York.

Proschan, F., and Serfling, R. J. (1974). "Reliability and Biometry." Ser Applied Math SIAM Philadelphia.

Reed, J. L., and Merrell, M. (1939). A short method for constructing an abridged life table. *Am. J. Hyg.* **30**:33–62.

Rees, P. H., and Wilson, A. G. (1973). Accounts and models for spatial demographic analysis I: Aggregate population. *Environ. Planning* **5**:61–90.

Rees, P. H., and Wilson, A. G. (1977). "Spatial Population Analysis." Arnold, London.

Regal, R. (1980). The F test with time-censored data. *Biometrika* **67**:479–481.

Reid, N. (1981a). Influence functions for censored data. *Ann. Stat.* **9**:78–92.

Reid, N. (1981b). Estimating the median survival time. *Biometrika* **68**:601–608.

Robinson, M. J., and Norman, A. P. (1975). Life tables for cystic fibrosis. *Arch. Dis. Child.* **50**:962–965.

Rockett, H., Antle, C., and Klimko, L. (1974). Maximum likelihood estimation with the Weibull model. *J. Am. Stat. Assoc.* **69**:246–249.

Rodriguez, G. (1984). The analysis of birth intervals using proportional hazard models. WFS/TECH No. 2314. World Fertility Survey, London.

Rodriguez, G., and Hobcraft, J. (1980). Illustrative analysis: Life table analysis of birth intervals in Columbia. "World Fertility Survey Scientific Reports," No. 16. International Statistical Institute, Voorberg, The Netherlands.

Rodriguez, G., Hobcraft, J., McDonald, J., Menken, J., and Trussell, J. (1983). A comparative analysis of the determinants of birth intervals. Comparative Studies, No. 30. International Statistical Institute, Voorberg, The Netherlands; and World Fertility Survey, London.

Roger, J. H., and Peacock, S. D. (1982). Fitting the scale as a GLIM parameter for Weibull, extreme value, logistic, and log-logistic regression models with censored data. *GLIM Newsl.* 6:30–37.

Rogers, A. (1966). The multiregional matrix growth operator and the stable interregional age structure. *Demography* 3:537–544.

Rogers, A. (1973). The multiregional life table. *J. Math. Sociol.* 3:127–137.

Rogers, A. (1975). "Introduction to Multiregional Mathematical Demography." Wiley, New York.

Rogers, A., ed. (1980). Essays in multistate demography. *Environ. Planning (A Special Issue)* 12(5).

Rogers, A. and Ledent, J. (1976). Increment–decrement life tables: A comment. *Demography* 13:287–290.

Rogers, A., and Willekens, F. J., eds. (1986). "Migration and Settlement: A multiregional Study." Reidel, Dordrecht, The Netherlands.

Rogers, A., Willekens, F. J., and Ledent, J. (1983). Migration and settlement: A multiregional comparative study. *Environ. Planning A* 15:1585–1612.

Sampford, M. R., and Taylor, J. (1959). Censored observations in randomized block experiments. *J. R. Stat. Soc. Ser. B* 21:214–237.

Sarhan, A. E., and Greenberg, B. G. (1962). "Contributions to Order Statistics." Wiley, New York.

SAS Institute (1982). "SAS User's Guide: Statistics." SAS Institute, Cary, North Carolina.

Saveland, W., and Glick, P. C. (1969). First marriage decrement tables by color and sex for the United States in 1958–1960. *Demography* 6:243–260.

Schmee, J., and Hahn, G. J. (1979). A simple method for regression analysis with censored data. *Technometrics* 21:416–432.

Schoen, R. (1975). Constructing increment–decrement life tables *Demography* 12:313–324.

Schoen, R. (1978). Calculating life tables by estimating Chiang's a from observed rates. *Demography* 15:625–635.

Schoen, R., and Baj, J. (1983). Marriage and divorce in five Western countries. Annual Meeting of the Population Association of America, Pittsburgh, Pennsylvania.

Schoen, R., and Land, K. (1979). A general algorithm for estimating a Markov-generated increment–decrement life table with applications to marital-status patterns. *J. Am. Stat. Assoc.* 74:761–776.

Schoen, R., and Nelson, V. E. (1974). Marriage, divorce, and mortality: A life table analysis. *Demography* 11:267–290.

Schoen, R., and Woodrow, K. (1984). Marriage and divorce in twentieth-century Belgian cohorts, *J. Family Hist.* 9:88–103.

Schoen, R., Urton, W., Woodrow, U. K., and Baj, T. (1985). Marriage and divorce in twentieth century American cohorts. *Demography* 22:101–114.

Schoenfield, D. (1980). Chi-squared goodness-of-fit tests for the proportional hazards regression model. *Biometrika* 67:145–153.

Seal, H. L. (1977). Studies in the history of probability and statistics, XXXV. Multiple decrements or competing risks. *Biometrika* 64:429–439.

Seigel, D. G. (1975). Life table rates and person month ratios as summary statistics for contraceptive trials. *J. Steroid Biochem.* 6:933–936.

Sen, P. K. (1981). The Cox regression model, invariance principles for some induced quantile processes and some repeated significance tests. *Ann. Stat.* 9:109–121.

Sewell, W. E. (1972). Life table analysis of the results of coronary surgery. *Chest* 61:481.

Shaked, M. (1977). Statistical inference for a class of life distributions. *Commun. Stat. A* 6:1323–1339.

Shapiro, S., Jones, E. W., and Densen, P. M. (1962). A life table of pregnancy terminations and correlated foetal loss. *Milbank Mem. Fund Q.* 40:7–45.

Sheps, M. C. (1961). Marriage and mortality. *Am. J. Public Health* 51:547–555.

Sheps, M. C. (1965). An analysis of reproductive patterns in an American isolate. *Popul. Stud.* 19:65–80.

Sheps, M. C., and Menken, J. A. (1973). "Mathematical Models of Conception and Birth." Univ. of Chicago Press, Chicago.

Shryock, H. S., and Seigel, J. S. (1973). "The Methods and Materials of Demography," Vols. I and II. U.S. Dept. of Commerce, Bureau of the Census, Washington, D.C.

Singh, S. N. (1964). On the time of first birth. *Sankhya* 26B:95–102.

Singh, S. N. (1968). A chance mechanism of variation in the number of births per couple. *J. Am. Stat. Assoc.* 63:209–213.

Singh, S. N., Bhattacharya, B. N., and Yadava, R. C. (1974). A parity dependent model for number of births and its application. *Sankhya* 36B:93–102.

Singh, S. N., Bhattacharya, B. N., and Yadava, R. C. (1979). An adjustment of a selection bias in postpartum amenorrhea from follow-up studies. *J. Am. Stat. Assoc.* 74:916–920.

Singhal, K. (1978). Topics in exponential regression models. Unpublished Ph.D. thesis, Univ. of Waterloo, Ontario, Canada.

Sirken, M. G. (1964). Comparisons of two methods of constructing abridged life table by reference to a 'standard' table. *Vital Health Stat. Ser.* 2(4):1–11. National Center for Health Statistics.

Smith, D. P. (1980). Life Table Analysis. World Fertility Survey, Technical Bulletin, No. 6, April. International Statistical Institute, Voorburg, The Netherlands, and World Fertility Survey, London.

Smith, S. (1979). Tables of working life for the United States, 1977: Substantive and methodological implications. Annual Meeting of the Population Association of America, Denver, Colorado.

Smith, S., and Horvath, F. (1984). New developments in multistate tables of working life. Annual Meeting of the Population Association of America, Minneapolis.

Snedecor, G. W., and Cochran, W. G. 1967. "Statistical Methods." Iowa Univ. Press, Ames, Iowa.

Spiegelman, M. (1968). "Introduction to Demography," rev. Ed. Harvard Univ. Press, Cambridge, Massachusetts.

Sprott, D. A. (1973). Normal likelihoods and relation to a larger sample theory of estimation. *Biometrika* 60:457–465.

Srinivasan, K. (1966). An application of a probability model to the study of interlive birth intervals. *Sankhya B* 28:175–192.

Stephens, M. A. (1978). Goodness of fit for the extreme value distribution. *Biometrika* 64:583–588.

Suchindran, C. M., Namboodiri, N. K., and West, K. (1979). Increment–decrement tables for human reproduction. *J. Biosocial Sci.* 11:443–456.

Sukhatme, P. V. (1937). Tests of significance for samples of the χ^2 population with two degrees of freedom, *Ann. Eugenics* 8:52–56.

Sundberg, R. (1974). Maximum likelihood theory for incomplete data from an exponential family. *Scand. J. Stat.* 1:49–58.

Sundberg, R. (1976). An iterative method for solution of the likelihood equations for incomplete data from exponential families. *Commun. Stat. B* 5:55–64.

Sverdrup, E. (1965). Estimates and test procedures in connection with stochastic models for deaths, recoveries and transfers between different states of health. *Scand. Actuar. J.* 48:184–211.

Tarone, R. E. (1975). Tests for trend in life table analysis. *Biometrika* 62:679–682.

Tarone, R. E., and Ware, J. (1977). On distribution-free tests for equality of survival distributions. *Biometrika* **64**:156–160.

Taylor, J. (1973). The analysis of designed experiments with censored observations. *Biometrics* **29**:35–43.

Taylor, W. F. (1964). On the methodology of measuring the probability of fetal death in a prospective study. *Hum. Biol.* **36**:86–103.

Temkin, N. R. (1978). An analysis for transient states with application to tumor shrinkage. *Biometrics* **34**:571–580.

Thiele, P. N. (1872). On a mathematical formula to express the rate of mortality throughout the whole of life. *J. Inst. Actuar.* **16**:213–239.

Thoman, D. R., and Bain, L. J. (1969). Two-sample tests in the Weibull distribution. *Technometrics* **11**:805–816.

Thoman, D. R., Bain, L. J., and Antle, C. E. (1970). Reliability and tolerance limits in the Weibull distribution. *Technometrics* **12**:363–371.

Thomas, D. C. (1981). General relative risk models for survival time and matched case-control analysis. *Biometrics* **37**:673–686.

Thomas, D. R. (1969). Conditional locally most powerful rank tests for the two-sample problem with arbitrarily censored data. Technical Report No. 7, Department of Statistics, Oregon State University.

Thomas, D. R., and Grunkemeier, G. L. (1975). Confidence interval estimation of survival probabilities for censored data. *J. Am. Stat. Assoc.* **70**:865–871.

Tietze, C., and Lewit, S. (1973). Recommended procedures for the statistical evaluation of intrauterine contraception. *Stud. Family Planning* **4**:35–41.

Tiku, M. L., (1967). Estimating the mean and standard deviation from a censored normal sample. *Biometrika* **54**:155–165.

Tolley, H. D., Burdick, D., Manton, K. G., and Stallard, E. (1978). A compartment model approach to the estimation of tumor increase and growth: Investigation of a model of cancer latency. *Biometrics* **34**:377–389.

Trussell, L., and Hammerslough, C. (1983). A hazards model analysis of the covariates of infant and child mortality in Sri Lanka. *Demography* **20**:1–26.

Trussell, J., and Menken, J. (1982). Life table analysis of contraceptive failures. *In* "The Role of Survey in the Analysis of Family Planning Programs" (A. Hermalin and B. Entwisle, eds.). International Union for the Scientific Study of Population, Liege.

Trussel, J., and Richards, Y. (1985). Correcting for unmeasured heterogeneity in hazard models using Heckman–Singer procedure. *In* "Sociological Methodology" (N. B. Tuma, ed.). Jossey Bass, San Francisco.

Trussell, J., Martin, L. G., Feldman, R., Palmore, J., Concepcion, M., and Abu Bakar, D. N. L. B. D. (1985). Determinants of birth-interval length in the Philippines, Malaysia, and Indonesia: A hazard-model analysis. *Demography* **22**:145–168.

Tsai, S. P., Lee, E. S., and Hardy, R. J. (1978). The effect of a reduction in leading causes of death: Potential gains in life expectancy. *Am. J. Public Health* **68**:228–233.

Tsiatis, A. A. (1975). A nonidentifiability aspect of the problem of competing risks. *Proc. Natl. Acad. Sci. U.S.A.* **72**:20–22.

Tsiatis, A. A. (1981). A large sample study of Cox's regression model. *Ann. Stat.* **9**:93–103.

Tsokos, C. P., and Shimi, I. N., eds. (1977). "The Theory and Application og Reliability, with Emphasis on Bayesian and Nonparamertric Methods," Vols. 1 and 2. Academic Press, New York.

Tuan, C. H. (1958). Reproductive histories of Chinese women in rural Taiwan. *Popul. Stud.* **12**:40–50.

Tukey, J. W. (1977). "Exploratory Data Analysis." Addison-Wesley, Reading, Massachusetts.

Tuma, N. B., and Hannan, M. T. (1984). "Soccial Dynamics, Models and Methods." Academic Press, Orlando, Florida.

Tuma, N. B., and Robins, P. K. (1980). A dynamic model of employment behavior. *Econometrica* **48**:1031–1052.

Tuma, N. B., Hannan, M. T., and Groeneveld, L. P. (1979). Dynamic analysis of event histories. *Am. J. Sociol.* **84**:820–854.

Turnbull, B. W. (1974). Nonparametric estimation of a survivorship function with doubly censored data. *J. Am. Stat. Assoc.* **69**:169–173.

Turnbull, B. W. (1976). The empirical disctribution function with arbitrarily grouped, censored and truncated data. *J. R. Stat. Soc. Ser. B.* **38**:290–295.

Turnbull, B. W., Brown, B. W., and Hu, M (1974). Survivorship analysis of heart transplant data. *J. Am. Stat. Assoc.* **69**:74–80.

United Nations (1955). "Age and Sex Patterns of Mortality: Model Life Tables for Under-developed Countries." United Nations, New York.

U.S. Department of Labor, Bureau of Labor Statistics (1982a). New work life estimates. Bulletin No. 2157 (by S. Smith). U.S. Govt. Printing Office, Washington, D.C.

U.S. Department of Labor, Bureau of Labor Statistics (1982b). The increment-decrement model. Bulletin No. 2135 (by S. Smith). U.S. Govt. Printing Office, Washington, D.C.

U.S. Department of Labor, Bureau of Labor Statistics (1950). Tables of working life: Length of working life for men. Bulletin No. 1001. U.S. Govt. Printing Office, Washington, D.C.

Vaeth, M. (1979). A note on the behavior of occurrence–exposure rates when the survival distribution is not exponential. *Scand. J. Stat.* **6**:77–80.

Vaupel, J. W., and Yashin, A. I. (1983). "The Deviant Dynamics of Death in Heterogeneous Populations." RR–83–1. International Institute for Applied System Analysis, Laxenberg, Austria.

Vaupel, J. W., and Yashin, A. I. (1985). Hetergeneity ruses: Some surprising effects of selection on population dynamics. *Am. Stat.* **39**:176–185.

Vaupel, J. W., Manten, K. G., and Stallard, E. (1979). The impact of heterogeneity in individual frailty on the dynamics of mortality. *Demography* **16**:439–454.

Ware, J. H., and Byar, D. P. (1979). Methods for the analysis of censored survival data. *In* "Perspectives in Biometrics" (R. M. Elashoff, ed.). Vol. 2. Academic Press, New York.

Weibull, W. (1951). A statistical distribution of wide applicability, *J. Appl. Mech.* **18**:293–297.

Weiss, K. M. (1973). A method for approximating age-specific fertility in the construction of life tables for anthropological populations. *Hum. Biol.* **45**:195–210.

Whitehead, J. (1980). Fitting Cox's regression model to survival data using GLIM. *Appl. Stat.* **29**:268–275.

Whittemore, A., and Altschuler, B. (1976). Lung cancer incidence in cigarette smokers: Further analysis of Doll and Hill's data for British physicians. *Biometrics* **32**:805–816.

Wilk, M. B., Gnanadesikan, R., and Huyett, M. J. (1962). Estimation of parameters of the gamma distribution using order statistics. *Biometrika* **49**:525–545.

Wilkin, J. C. (1981). Recent trends in the mortality of the aged. *Trans. Soc. Actuar.* **33**:11–44.

Willekens, F. J. (1980). Multistate analysis: Tables of working life. *Environ. Planning A* **12**:563–588.

Willekens, F. J. (1982). Multidimensional population analysis with incomplete data. *In* "Multidimensional Mathematical Demography" (K. Land and A. Rogers, eds.). Academic Press, New York.

Willekens, F. J. (1985a). The marital status life table. *In* "Family Demography: Methods and their Applications" (J. Bongaarts, T. K. Burch, and K. W. Wachter, eds.). Oxford Univ. Press, London and New York.

Willekens, F. J. (1985b). Multiregional demography, Working paper No. 59. Netherlands Interuniversity Demographic Institute, Voorburg.

Willekens, F. J., and Drewe, P. (1984). A multiregional model for regional demographic projection. *In* "Demographic Research and Spatial Policy" (H. ter Heide and F. J. Willekens, eds.). Academic Press, London.

Willekens, F. J., Sha, I., Sha, J. M., and Ramachandran, P. (1982). Multistate analysis of marital status life tables. Theory and application. *Popul. Stud.* **36**:129–144.

Williams, J. S. (1978). Efficient analysis of Weibull survived data from experiments on heterogeneous patient populations. *Biometrics* **34**:209–222.

Wilson, E. G. (1954). The standard deviation for sampling for life expectancy. *J. Am. Stat. Assoc.* **33**:705–708.

Wolfbein, S. L. (1949). The length of working life. *Popul. Stud.* **3**:286–294.

Wolfers, D. (1968). Determinants of birth intervals and their means. *Popul. Stud.* **22**:253–262.

Wolynetz, M. S. (1979). Statistical algorithms AS138 and AS139. *Appl. Stat.* **28**:185–206.

Woodbury, M. A., and Manton, K. G. (1977). A random walk model of human mortality and aging. *Theor. Popul. Biol.* **11**:37–48.

Woodbury, M. A., Manton, K. G., and Stallard, E. (1979). Longitudinal analysis of the dynamics and risks of coronary heart disease in the Farmingham study. *Biometrics* **35**:575–585.

Woodbury, M. A., Manton, K. G., and Stallard, E. (1981). A dynamic analysis of chronic disease development: A study of sex specific changes in coronary heart disease incidence and risk factors in Framingham. *Int. J. Epidemiol.* **19**:355–366.

Zelen, M. (1959). Factorial experiments in life testing. *Technometrics* **1**:269–288.

Zelen, M. (1960). Analysis of two-factor classifications with respect to life tests. *In* "Contributions to Probability and Statistics" (I. Olkin, ed.). Stanford Univ. Press, Stanford, California.

Zelen, M. (1966). Applications of exponential models to problems in cancer research. *J. R. Stat. Soc. Ser. A* **129**:368–398.

Zippin, C., and Armitage, P. (1966). Use of concomitant variables and incomplete survival information in the estimation of an exponential survival parameter. *Biometrics* **22**:665–672.

Author Index

Subject Index